NATURAL SCIENCE AND THE ORIGINS OF
THE BRITISH EMPIRE

EMPIRES IN PERSPECTIVE

Series Editors:	Tony Ballantyne
	Duncan Bell
	Francisco Bethencourt
	Caroline Elkins
	Durba Ghosh
Advisory Editor:	Masaie Matsumura

TITLES IN THIS SERIES

Between Empire and Revolution: A Life of Sidney Bunting, 1873–1936
Allison Drew

A Wider Patriotism: Alfred Milner and the British Empire
J. Lee Thompson

Missionary Education and Empire in Late Colonial India, 1860–1920
Hayden J. A. Bellenoit

Transoceanic Radical, William Duane: National Identity and Empire 1760–1835
Nigel Ken Little

FORTHCOMING TITLES

Empire of Political Thought: Indigenous Australians
and the Language of Colonial Government
Bruce Buchan

Ireland and Empire, 1692–1770
Charles Ivar McGrath

The English Empire in America, 1602–1658: Beyond Jamestown
Louis H. Roper

NATURAL SCIENCE AND THE ORIGINS OF THE BRITISH EMPIRE

BY

Sarah Irving

Routledge
Taylor & Francis Group

LONDON AND NEW YORK

First published 2008 by Pickering & Chatto (Publishers) Limited

Published 2016 by Routledge
2 Park Square, Milton Park, Abingdon, Oxfordshire OX14 4RN
711 Third Avenue, New York, NY 10017, USA

First issued in paperback 2015

Routledge is an imprint of the Taylor & Francis Group, an informa business

BRITISH LIBRARY CATALOGUING IN PUBLICATION DATA

Irving, Sarah
Natural science and the origins of the British Empire. – (Empires in perspective)
1. Natural history – History – 17th century 2. Science – History – 17th century
3. Imperialism and science 4. Great Britain – Colonies
I. Title
509.1'71241'09032

ISBN-13: 978-1-138-66522-4 (pbk)
ISBN-13: 978-1-8519-6889-3 (hbk)

Typeset by Pickering & Chatto (Publishers) Limited

CONTENTS

ACKNOWLEDGEMENTS

This book is in some sense a product of the enduring legacy of the former British Empire.

As an Australian studying at Cambridge University on a British Commonwealth Scholarship, it was perhaps fitting that I chose to write on an imperial theme. The debts of gratitude I have accumulated during my three years at Cambridge and one year at Oxford are also testament to a post-imperial world. I was sustained by my family in Sydney, Australia, supported by colleagues and mentors in England, and inspired by my friends around the Atlantic.

The Association of Commonwealth Universities made it possible for me to complete my PhD at King's College, Cambridge, on a British Commonwealth Scholarship. I want to thank King's College, its students and Fellows, for allowing me to live and work among them. I wrote the book manuscript during my term as a Junior Research Fellow at Wolfson College, Oxford. I relished the stimulating and democratic atmosphere of the college and I would like to extend my sincere thanks to the Fellows and students of Wolfson.

For their financial assistance in the form of grants or scholarships, I would like to thank King's College, Cambridge, Wolfson College, Oxford, the Association of Commonwealth Universities and Warwick University. The book was awarded the Royal Society of Literature and Jerwood Foundation Award for Non-Fiction. It is with deep gratitude and honour that I thank these two institutions for their support.

A number of friends and colleagues read different chapters of my PhD and gave me numerous valuable suggestions. For this, and for their generosity, I would like to thank Conal Condren, John Dunn, Andrew Fitzmaurice, David Sacks and Simon Schaffer. I would particularly like to thank Simon for a number of discussions which developed and clarified my thinking about early modern natural philosophy. Andrew Fitzmaurice and Conal Condren introduced me to the history of political thought and have encouraged me in my academic career since I was an undergraduate. I am ever grateful. I would particularly like to thank Peter Harrison and Richard Serjeantson for examining the PhD dissertation upon which this book is based and for giving me helpful suggestions for revising the thesis for publication. I could not have asked for more astute examiners. Throughout the time that it has taken to write the dissertation and the book, Michael Middeke and his colleagues at Pickering & Chatto have been patient and forthcoming with advice.

I owe special thanks to my close friends who have supported me and shared with me their lives, even through the tyranny of distance. Each of them inspires me. David Blunt, John William Devine, Jacqui Mills and above all Hans Wietzke have enriched my life.

To Mark Goldie I owe a debt of gratitude I can never fully repay. I cannot imagine a more meticulous reader of my work, nor a reader whose sharp bibliographic mind could point me towards so many sources and potential critiques of my arguments. My work has improved and matured immeasurably under Mark's guidance and supervision. For all this, and for his generosity with his time, his wisdom and his ongoing encouragement, I want to offer Mark my fullest thanks.

My greatest debt is to my family. My parents, Terry and Sue Irving, my brother Nick Irving, and my grandmother Norma Chapman have supported and encouraged me in more ways than I can countenance, and for so many years: across continents, both hemispheres, and the former British Empire. I dedicate this book to them. Finally, I would like to thank my father for introducing me to the study of history.

Oxford, December 2007

For my family

PREFACE

Empire is the clash of bodies, weapons and ideas. When the rock musician Neil Young described a 'white boat coming up the river' in his arresting song 'Powderfinger', he could have been writing a metaphor for British imperialism. Armed with a gun and flying a flag, the boat's arrival heralds not negotiation but conquest. From the marshy coast of the Chesapeake to the rocky inlets of Sydney Harbour, the invasion of the white boat marked a new epoch in world history. Empires shaped the modern world and they still animate our political imagination.

Even in a post-colonial era we remain acutely conscious of the phenomenon of empire. Imperialism is one of the most prominent and contested concepts in understanding contemporary international politics. This fact contributes to its complexity, but also to its intellectual appeal. If, as Sankar Muthu wrote, one of the reasons we study the history of political thought is to 'gain the perspective of another set of assumptions and arguments that are shaped by different historical sensibilities'[1] then there are few more compelling subjects than empire.

This book investigates how the idea of the British Empire came into being. What constellation of events and intellectual manoeuvres produced the conception of the British Empire that we hold today? Such a discussion of ideas might sound esoteric, even arcane. Intellectual historians are often accused of neglecting the material conditions of people's daily lives. This is a book about ideas, certainly. But it is not a book about ideas that had no influence upon the experiences of men and women. The subject of this study is a tradition of empire that helped found the theory of property that legitimized the transplantation – and dispossession – of people throughout the British colonies.

My project is to investigate a central part of the ideological origins of the British Empire. I want to understand the connection between an idea of empire as man's plenary dominion over the earth in the Garden of Eden, and the early English colonization of the Atlantic world. In the early modern period, the idea of Adam in Eden had a strong hold on European consciousness, a fact easily overlooked when we view the past from our secular, contemporary perspective. It is often thought that modern science was founded upon a rejection of religion, but in fact the opposite is the case. Modern science finds its most important forebears in the 'new philosophy' that emerged in seventeenth-century Europe. Its practitioners were adamant that the purpose of studying nature was to recover man's original dominion over the earth, bestowed upon Adam in Eden but lost in the Fall. As Francis Bacon implored readers of his iconoclastic *New Organon*, 'Just let man recover the right over nature which belongs to him by God's gift, and give it scope; right reason and sound religion will govern its use.'[2]

The exponents of the new natural philosophy adopted Bacon's aim. Robert Boyle, Robert Hooke, John Locke and other founding members of the Royal Society of London aimed to use natural philosophy to rebuild man's original dominion over nature. Robert Boyle, for example, often quoted the Genesis verse where God enjoins Adam to 'be fertile and increase, fill the earth and master it; and have an empire over the fish of the sea, the birds of the sky, and all the living things that creep on earth' (Genesis 1:28). Natural philosophy, Boyle believed, should aim to recover man's lost dominion over nature.

What is fascinating about this ideal of subduing the earth is that it was conceptualized as an *empire*. Did natural philosophers' appeals to the restoration of man's original empire over nature have anything to do with their involvement in the institutions of *colonial* empire-building? Was there a relationship between these two traditions of empire: the Biblical tradition of man's dominion over nature, and that of England over her colonies?

This book argues that colonization became central to the natural philosophical project of recovering man's empire over nature. There were two ways in which colonies were significant to this enterprise. First, they were seen as a repository of information about the natural world. Before the Fall, Adam possessed an encyclopedic knowledge of nature which enabled him to give all the world's creatures their correct names. Because this knowledge was lost in the Fall, when the English encountered the Americas they believed their natural environment contained the plants and animals which Adam had once known. It was, in short, a new Eden. Recovering man's encyclopedic knowledge of nature, therefore, was part of a religious project.

The first significance of colonies, then, was epistemic. The second was agrarian. Colonies were seen as land ready to be fenced off, cultivated and made into private property. The other way man would rebuild his foregone empire was through agrarian labour. England, on behalf of mankind, would make the earth fruitful again. Man should utilize the resources of the world because as John Locke put it, quoting Paul's First Letter to Timothy, 'God gave us all things richly to enjoy.'[3] An ideal of cultivation, based upon the redemptive possibility of labouring to amend the agrarian implications of the Fall, was developed throughout the seventeenth century. Eventually, I argue, it became the basis of Locke's theory of property. Locke argued that the land which we cultivate – which we 'subdue' to use his Biblical idiom – could be legitimately taken as private property. Locke's theory had implications for the Atlantic colonies. If indigenous people did not subdue the earth and cultivate it then the land was free to be taken by the British.

Beginning in the seventeenth century, I argue, England's colonial empire became tied to the redemptive project of restoring man's prelapsarian empire over nature. This book is a history of the Biblical tradition that, in subsequent centuries, the British would use to help justify the dispossession of indigenous people from Australia to North America. I hope it helps us understand how the British legitimized their white boat coming up the river.

INTRODUCTION

Rarely do historians describe seventeenth-century English natural philosophers as colonialists. Rarely, too, has there been serious scholarly discussion of natural philosophers' involvement in the intellectual and practical processes of English empire-building. This is a curious omission. It may surprise some historians of the British Empire to learn that Robert Boyle served on the board of the English East India Company, held shares in the Hudson's Bay Company, and served as President of the New England Company, a missionary society which sponsored the translation of the Bible into the indigenous Algonquian language. He also served on the Council for Foreign Plantations. Boyle's intellectual predecessor Francis Bacon held shares in the Virginia Company, the Newfoundland Company, and wrote extensively on the issue of colonizing Ireland. Moreover, many of Boyle's contemporaries in the Hartlib Circle, as well as those in the Royal Society of London such as Sir Hans Sloane and Henry Oldenburg, maintained an avid interest in the English colonies.

Natural philosophers' colonial involvement, and their absence from histories of the British Empire's intellectual origins, compels a historical question. What understanding of empire did natural philosophers hold? The answer to this question is startling. In his tract, *Of the Usefulness of Experimental Natural Philosophy*, Boyle described his overarching project as the re-creation of '*The Empire of Man over inferior Creatures*'.[1] This was the framework through which he understood his involvement in the institutions of English colonization.

Boyle's motivation for involving himself in English colonization was not that of establishing virtuous new commonwealths in North America. Nor was it primarily about the profit from the plantations in the Caribbean or in Ireland, the latter of which he was himself a direct beneficiary. It was not even England, as a realm under the unchallenged authority of King Charles II, which constituted Boyle's primary ideal of 'empire'. Rather, the 'empire of man over inferior creatures' denoted the original dominion that Adam commanded over nature in the Garden of Eden.

British Empire scholarship has largely focused on the emergence of the British Empire as a territorial entity and upon its inheritance from the Roman law tradition in which the primary conception of empire was of the un-impinged sovereignty of the monarch over a composite collection of territories. According to this account, the British Empire's primary intellectual inheritance is in Roman law, and its texts concern English state-building. When confronted with this story, Boyle's conception of empire as man's prelapsarian dominion over the earth seems incongruous: its inheritance is not Roman but theological, and its texts concerned not state-building but the Old Testament Creation narrative.

This narrative, articulated in the book of Genesis, tells a markedly different story about empire. In the Garden of Eden, God endowed Adam with a mastery over the natural world. 'Be fruitful and multiply, and fill the earth and subdue it; and have dominion over the fish of the sea and over the birds of the air and over every living thing that moves upon the earth' (Genesis 1:28). Adam's mastery over the earth was partly manifested in his perfect knowledge of the natural world; knowledge that enabled him to give correct names to creatures. In the Fall, however, Adam, who stands as a cipher for mankind, lost both his knowledge and his empire over nature.

To historians of science, this idea of empire – at least as it pertained to man's proper relationship to the earth – is well known as the foundation of Francis Bacon's project *The Great Instauration* which aimed to 'enlarge the bounds of humane empire to the effecting of all things possible'.[2] Bacon's aim was adopted by the exponents of the new natural philosophy, during what might, in an historically messy shorthand, be called the Scientific Revolution. Robert Boyle, John Locke and other founding members of the Royal Society of London aimed to use natural philosophy to restore man's original dominion over nature. The fact that the core of this ideal of man's dominion is actually a theory of *empire* remains unexplored. Did natural philosophers' appeals to the restoration of man's *original* empire over the world have anything to do with their involvement in the institutions of *colonial* empire-building? What was the relationship between the concepts of geographic empire, and man's prelapsarian empire?

This book has two aims: the first is historical and the second is methodological. The historical aim is to use the tradition of empire in the work of English natural philosophers from Francis Bacon (1561–1626) to John Locke (1632–1704) as the basis of an intellectual history of the origins of the British Empire. This is an interdisciplinary project which takes the form of bringing natural philosophy back into the history of the British Empire and, conversely, bringing the concept and origins of the British Empire to bear on early modern English science.

My methodological aim concerns the usefulness of political languages as a means of understanding early modern intellectual transmission. In intellectual history, the dominant way of understanding the process of intellectual transmission is through a series of political vocabularies or languages: that is, through a set of terms or keywords that constitute a kind of rhetorical tool box; a set of resources into which authors consciously delve in the context of a political debate. I hope to demonstrate, however, that the rubric of 'political' languages is inadequate for understanding the intellectual origins of the British Empire. Languages are certainly the means through which early modern ideas were articulated, but the characterization of such languages as *political* is problematic. There are two reasons for this. First, the notion that early modern vocabularies are best described as 'political' does not allow sufficient room for natural philosophy or theology. It is too secular a context. Natural philosophers' use of the language of Adam's empire did not primarily occur in political debate; the language is better understood as theological, or natural philosophical, than political. The characterization of early modern languages as 'political' misleads us into assuming a secular context for the generation of these ideas. The seventeenth-century mental universe – in particular the mindset of natural philosophers – was deeply theological.

The second point is that, associated with the rubric of *political* vocabularies is the idea that the most formative intellectual tradition in early modern England was that derived from Roman thought. In the context of the intellectual origins of the British Empire, however, the Bible was just as influential a text as the work of Cicero. My methodological aim is to extend the conception of early modern vocabularies beyond the political. I will introduce a theological and natural philosophical vocabulary of Adam's empire to the group of languages which shaped the intellectual life of the seventeenth-century Atlantic world. In doing so, I aim to broaden our understanding of the intellectual origins of the British Empire beyond both the Puritan and the Roman inheritances to include the formative traditions of theology and natural philosophy.

The book pursues these aims by mounting two central arguments. The first concerns the importance of the New World, and of a concept of 'empire', to early modern English natural philosophers. The reason why English natural philosophers were so interested in the New World, and particularly in English colonies, was because they held a theory of empire as man's original dominion over nature. It was the emphasis upon restoring man's original encyclopedic knowledge of the natural world which generated their interest in the New World, and ultimately in its English colonies. Put another way, there existed a concept of empire – largely overlooked by British Empire scholars – which influenced a tradition of natural philosophers involved in the institutions of empire-building.

One might question why I choose to focus on the New World and not, for example, the trading posts being established on the Indian subcontinent during this period. The New World was of particular interest to natural philosophers because it was unknown to the ancients. The natural knowledge it would yield, therefore, was assumed to be that which was lost at the moment of the Fall. It would be misguided and nearly impossible, however, to study natural philosophical attitudes toward the New World without extensive references to Ireland, so the scope of this book is Atlantic. This is not to say that other overseas places, such as the Indian subcontinent, were not of interest to my subjects, but rather that for considerations of space, this is a book about Atlantic history.

My second argument concerns the role that natural philosophy played in the ideological origins of the British Empire. The idea of man's dominion over nature became part of the basis for the Lockean theory of property. In subsequent centuries, the British used this theory to help justify their colonial possessions, whatever Locke's original intention. The final chapter will demonstrate that Locke's theory of property in the *Two Treatises of Government* was rooted in the ideal of restoring man's prelapsarian empire over the earth. This was a project of redemptive labour in which man had to cultivate the soil in order to return the earth to its fruitfulness and man to his proper position of earthly dominion. As Locke put it, 'Man had a right to a use of the Creatures, by the Will and Grant of God'.[3]

Although this argument rests ultimately upon my analysis of Locke, its groundwork is laid in the preceding chapters, which trace what I argue were two separate but related aspects of the restoration of man's plenary empire over the earth: one agrarian and the other epistemological. The agrarian aspect consisted in the idea that man must cultivate the soil and in doing so return the earth to its original fruitfulness. After all, the earth also suffered because of the Fall. 'Cursed is the ground for your sake; In toil you shall eat of it all the days of your life' (Genesis 3:17). An ideal of cultivation based upon the redemptive possibility of labouring to amend the agrarian implications of the Fall was developed throughout the seventeenth century and eventually became the basis of Locke's theory of property.

The second aspect of Adamic empire was epistemic. This was the recovery of man's encyclopedic knowledge of the natural world. The idea of recovering and collating knowledge was not unique to the seventeenth century. As Richard Yeo has shown, 'Protestant theologians in particular, were exercised by the hope of restoring former pristine knowledge of nature, languages and morality',[4] and that, in the mind of Bacon, for example, the 'search for natural knowledge [was] a precondition for spiritual redemption'.[5] Aside from Bacon and a number of the members of the Royal Society, whose primary concern was the recovery of man's epistemic empire, the natural philosophers I explore were predominantly interested in both aspects of the recovery of man's original empire over nature. Over the course of the century, however, there was a gradual shift towards an emphasis upon the project of cultivating the earth and, ultimately in Locke's work, a connection between this agrarian project and English colonization.

Historians and the British Empire

Approaching the British Empire as an intellectual historian is an odd experience. One expects to find histories of the *idea* of empire, and of the ideologies of empire, perhaps along the line of the kind of 'history of concepts' which has taken hold of other key political ideas like democracy. For this kind of study, one would think that 'empire' would be a marvellous candidate. Yet scholars of the British Empire have exhibited a certain resistance to writing about their subject's conceptual origins. There are, I think, two reasons for this.

The first has to do with the nature of the history of political thought. There is a certain incongruity when the idea of empire is placed in the context of other political concepts: empires transcend the 'state', they transcend the 'nation', and so, it is often thought, they transcend the territory of the historian of political thought, insofar as politics is an activity of citizens, subjects and the State.[6] A good deal of British Empire scholarship, however, deals with the theory of *imperialism*. This is partly because of the importance of the concept of imperialism to the Left, for example in the works of Vladimir Lenin[7] and Karl Kautsky[8] and more recently Edward Said.[9]

But in these studies, empire becomes an *ism*; an abstract characteristic of the capitalist economic structure which defines modernity. There is little space in this debate for a discussion of the sixteenth- and seventeenth-century intellectual origins of the British Empire, in which ideas of profit and capital were less important than those of virtue or Old Testament theology, for example. It is not surprising that the etymology of the term 'imperialism' presented by Raymond Williams in his famous study *Keywords* (1976) sees the term as part of a *modern* political lexicon. The word imperialism is, he states, a nineteenth-century invention.[10] And so it is. But *empire* was not. Williams's choice of 'imperialism' as his keyword is perhaps the best illustration of the fact that, among many scholars, *imperialism* rather than *empire* is the issue.

The second reason that there has been little scholarly investigation of the conceptual origins of the British Empire concerns the division of labour among historians. In recent years the British Empire has become the subject of new types of history-writing: post-colonialism and cultural history have reoriented the scholarly debate towards the idea of cultural encounters between colonizers and colonized. Most of the primary sources for this theme in British history pertain to the eighteenth and nineteenth centuries, with the spread of English colonization into India, Africa and the south Pacific. Related to this is the fact that the story of the rise and fall of the British Empire has largely been told as the story of the rise and fall of the 'second' British Empire, that which began in India and spread to Africa and Australasia in the latter half of the eighteenth century, and was founded upon military conquest and the exploitation of indigenous people. Consequently, the haphazard colonization in the Atlantic world, that amorphous outgrowth of early modern English state-building, has been omitted from many histories which dealt only with the second empire.

In Eric Hobsbawm's series of general histories, *The Age of Empire* denotes the period from 1875 to 1915. The British Empire's origins in the late sixteenth-century colonization of Ireland and tentative colonization of North America are implicitly out of synch with Hobsbawm's epoch-defining nineteenth- and twentieth-century empires. In European scholarly consciousness, the British Empire is a capitalist phenomenon and its history is seen as distinct from Britain's domestic history. There exists what David Armitage has called the 'continuing disjuncture between "British" and "Imperial" histories'.[11]

Until recently, Puritanism was the focus of much scholarship on the history of English colonization.[12] Although I do not deal explicitly with Puritanism, my case for the importance of theology to our understanding of the intellectual origins of the British Empire echoes this scholarship's emphasis upon the religious ideals of the early colonizers. The immediate historiographic context for this book is the recent shift away from the scholarship on Puritanism and towards a more secular reading of the British Empire's ideological origins in which the traditions of Roman law and civic humanism are central. This move was pioneered by David Armitage's *Ideological Origins of the British Empire* (2000).

According to Armitage, the intellectual inheritance of the British Empire is to be found in the Roman law tradition, in which the term 'imperium' denoted the un-impinged, indivisible sovereignty of the monarch over a number of territories. In England's case, this was a composite monarchy comprised of territories in the British Isles. Anglo-Saxon kings, for example, claimed to be rulers of an *Imperium Anglorum*. As Frances Yates showed, under the reign of Elizabeth I, an imperial idea emerged which concerned religion above all. It was an idea of 'sacred empire' characterized by royal supremacy over both church and state and was articulated in works such as John Dee's *General and Rare Memorials Pertayning to the Perfect Arte of Navigation*. In Dee's work, 'the tale of the lands and seas to which she can lay claim is based both on the dominions mythically reported to have been held by the British King Arthur and on those over which the Saxon King Edgar ruled'.[13]

The concept of a British Empire, however, emerged in the context of Anglo-Scottish relations in the 1540s and drew upon the idea of empire in Roman law: *Rex in regno suo est imperator* – the king is an emperor within his own kingdom.[14] The 1540s saw the birth of the concept of the 'empire of Great Britain', and it also saw the beginnings of a Protestant conception of that empire.[15] Authors of pamphlets on behalf of Henry VIII advocated a union of England and Scotland under English supremacy, claiming that the archipelago of Britain was his empire in the tradition of Brutus. It was an empire in the sense that the monarch possessed unlimited authority within his territory, and at the same time he was independent from the Pope. Thus the three kingdoms of England, Scotland and Ireland were united in the form of a composite monarchy, in which all three were independent, but under one sovereign. As Armitage shows, the Roman lineage of empire as unchallenged sovereignty 'encouraged the fastening of territorial boundaries'.[16] This was a concept of empire as sovereignty which was exercised geographically.

The fact that the emergence of a concept of empire in England took place in the context of debates about the nature of the three kingdoms illuminates one of the underlying themes of Armitage's study, which is that state-building and empire-building went hand in hand. The other historical moment of the creation of the British Empire also involved a state-building exercise. This was the colonization of Ireland which, as Armitage has shown, was based upon a Roman model. Ironic from the standpoint of history, in the Irish context colonization was put forward as the solution to the troubles of governing the area and keeping it under English control. It was these acts of state-based consolidation, in Scotland and in Ireland, that Armitage argues formed the basis of the British Empire.

This process was, however, imperfect and contingent.[17] The Irish and the Scottish did not think of themselves as British, and the Scottish pursued colonization on the other side of the Atlantic under their own identity and auspices at Darien on the Isthmus of Panama, much to the chagrin of the English.[18] Moreover, as Armitage has shown, the term 'British Empire', denoting a single political body encompassing the Atlantic colonies as well as the British archipelago, was not used until the late seventeenth century at the earliest.[19]

Caution is needed, therefore, when discussing the intellectual origins of the British Empire. It is important to remember that despite the fact that the concept of the British Empire emerged at the end of the seventeenth century, its ideological origins were shaped much earlier. We can legitimately speak of the influence of seventeenth-century natural philosophy upon the intellectual formation of the 'British Empire'. The reason for this is that there is a foundational connection between the idea of man's plenary empire and the Lockean theory of property which, whatever Locke's original intention, ultimately became central to the British Empire's ideological apparatus. For the purposes of historical accuracy, I will refer to 'English' colonization and empire-building throughout the book.

A central problem in the ideological origins of the British Empire, David Armitage argues, was the uneasy coexistence of *dominium* (property, or the rights that landlords had over their estates) and *imperium* (the unlimited, independent authority of a monarch over his territory which did not necessarily encourage expansion). This problem is the basis of one of Armitage's most important claims: that the British ideologies of empire were never able to unite a theory of sovereignty with a theory of property. This, he argues, was the 'ultimately combustible dilemma at the core of British Imperial Ideology'.[20] According to Armitage, there were several attempts at incorporating *dominium* and *imperium* but the disjunction remained, and it explains why Thomas Jefferson and John Adams could show that, since the first discovery of America had been undertaken by individuals or companies rather than by the state, Parliament could make no claims over the American colonists who were encompassed only by royal *imperium*.[21]

I disagree with the claim that there was a disjunction between the concepts of *imperium* and *dominium* – sovereignty and property – and that the two ideas were never properly brought together. I aim to demonstrate that in the Biblical tradition of empire, these two ideas existed in harmony. According to the Hartlib Circle, and later John Locke, for example, dominion over land would lead to the restoration of man's earthly sovereignty. Furthermore, the terms *imperium* and *dominium*, and their English translations empire and dominion, were often used interchangeably as synonyms.

Although this argument will be explicated later, we can demonstrate the point briefly by comparing the King James Bible's use of the terms with that of Francis Bacon in his 'Temporis Partus Masculus', the Masculine Birth of Time. The full title of this text is 'Temporis Partus Masculus Sive Instauratio Magna Imperii Humani in Universum',[22] which can be translated as 'The Masculine Birth of Time or, The Great Instauration of the Empire of Man Over the Universe'. Bacon chose to use the term *imperium* to describe man's original power over the world. According to Armitage's argument, however, we would expect Bacon to have used the term *dominium* rather than *imperium*, because 'dominion' was the word used in the book of Genesis to describe man's plenary empire over nature.[23] Given that Bacon was referring to this theological idea in the book of Genesis, there are two possible interpretations for Bacon's word choice. Either Bacon was intentionally altering the meaning of the passages in Genesis because he believed that man's plenary 'dominion' and man's plenary 'empire' were different ideas, or the terms were, in this context, synonyms. The latter is clearly the case. This linguistic ambiguity between the concepts of dominion and empire, a phenomenon that we witness throughout the seventeenth century, challenges the argument that the two were irretrievably disjointed.

For Armitage, Genesis 1:28 was a scriptural basis for colonization and it was also the Biblical justification for the agricultural argument for property rights. This was the idea that, as Locke put it, if we mix our labour with the earth it becomes legitimately ours. For Armitage, however, this is only a theory of property – of *dominium* – rather than empire. By contrast, I want to suggest that the Biblical injunction to improve the earth was in fact attached to the idea of Adam's original empire over nature. As I show in the final chapter on John Locke, the agricultural argument for property was part of the Adamic *imperial* project because recovering the fruitfulness of the earth will give man an empire over it. Locke was making an argument not only about property but also about man's original *empire*. Because Locke employed a theory of property that encompassed a theory of empire, I hope to demonstrate, contra Armitage, that *dominium* and *imperium* coexisted in the ideological origins of the British Empire.

The emphasis upon the Roman lineage of empire is extremely valuable to our understanding of the British Empire's ideological origins. It is also a very important depiction of the role of debates about early English state-building in shaping claims to empire. The point of this book is not to suggest that the conception of man's plenary empire over nature was opposed to, or in competition with, the Roman tradition. Rather, my point is that the British Empire has a largely overlooked intellectual lineage in natural philosophers' adoption of a theology centred upon Adam.

A second recent study to bring the legacy of classical thought back into the intellectual history of English colonization is Andrew Fitzmaurice's *Humanism and America: An Intellectual History of English Colonisation 1500–1625*. Fitzmaurice argues that the 'humanist imagination dominated colonizing projects'[24] and that early English colonists aimed to establish new commonwealths. They were often anxious about dispossessing the indigenous inhabitants of the land, and about the potential for moral corruption on the colonial periphery. Francis Bacon, for example, displayed a civic humanist anxiety about 'displanting' native peoples. This is an astute observation. I aim to show that the rest of Bacon's work reveals him not just as a civic humanist but also as a natural philosopher for whom the advancement of learning would enable man to recover his original empire over nature. The New World was vital to this epistemic imperial project.

There are several recent scholarly trends in the broad field of British Empire historiography. In general, these have accompanied the increasing interdisciplinarity in the humanities and the attention to cultural history. The first trend decentres Britain as the subject of history and places it in the broader context of the Atlantic world, the Commonwealth and the kingdoms of Ireland and Scotland. The creation of the British Empire is now frequently seen in the context of the development of England's sixteenth- and seventeenth-century colonial policies toward Ireland, and her relationship with Scotland. The 'New British History' as it is called, includes for example the new five-volume *Oxford History of the British Empire*, and recent work on the relationship between the English colonization of Ireland and that of America includes the work of Nicholas Canny and David Beers Quinn.[25]

The relatively new field of Atlantic history has emerged from a recognition of the interrelationship of the Three Kingdoms and North America in the formation of the early modern world and its intellectual traditions. Atlantic history tells 'the story of the creation, destruction and recreation of communities as a result of the movement, across and around the Atlantic basin, of people, commodities, cultural practices and ideas' as J. H Elliott puts it.[26] This is part of a larger trend to examine the way that cultural encounters shaped the mental categories of their European colonizers.[27]

The scholarship on Atlantic history would benefit from an exploration of the role that natural philosophy played in shaping the concept of empire. It is especially indicative that the new five-volume *Oxford History of the British Empire* does not deal with the influence of natural philosophy on the conceptual origins of the Empire. The first volume, on the 'Origins of Empire', makes no reference to natural philosophy or science. The second volume on the eighteenth century does include Richard Drayton's chapter, 'Knowledge and Empire',[28] and the fifth volume includes another chapter by Drayton on 'Science, Medicine and Empire',[29] but both deal primarily with the period after the seventeenth century. Moreover, as Drayton's analysis and synopsis of the historiography reveals, the recent burgeoning interest in the role that science played in British empire-building has seen no study of the *concept* of empire.[30]

Nevertheless, a number of illuminating studies focus upon the role that science played in the English imperial endeavours and early American colonial history. Joyce Chaplin's *Subject Matter: Technology, the Body and Science on the Anglo-American Frontier, 1500–1676*,[31] deals with the relationship between science and English colonization in the context of early American history with particular attention to the history of race and disease. Richard Grove's *Ecology, Climate and Empire: Colonialism and Global Environmental History*[32] and his *Green Imperialism: Colonial Expansion, Tropical Island Edens and the Origins of Environmentalism, 1600–1860*[33] take the phenomenon of empire as the defining feature in the history of ecology. Grove charts the origins of environmental concern and argues that 'the seeds of modern conservationism developed as an integral part of the European encounter with the tropics and with local classifications and interpretations of the natural world and its symbolism'.[34] For natural philosophers in the early modern period, for example, 'the physical environment began to acquire the attributes of religious experience and purpose ... Paradise had become a realizable geographical reality, or so it seemed.'[35]

In the British context, Richard Drayton engages in a similar project in *Nature's Government: Science, Imperial Britain and the Improvement of the World* (2000).[36] This is one of few studies to bring together the history of science with the intellectual history of the British Empire. Underpinning Drayton's work is the argument that 'ideas of Providence, and of Adamic responsibilities and prerogatives, were the ideological taproot of the First British Empire and, translated into political economy, they underpinned the Second'.[37] In other words, there was a Protestant theological background to the ideals of economic improvement which animated the British Empire from the late seventeenth century onwards.

I hope to develop this attention to the theological, intellectual lineage of the British Empire by arguing that Adamic ideas of man's dominion over the world constituted a theory of empire. This Biblical understanding of empire warrants being conceptualized as an intellectual tradition of empire, such that we can map its use and development by natural philosophers from the late sixteenth to the early eighteenth centuries.

Aside from these important works, the bulk of the literature on the relationship between science and the British Empire falls into two categories. First, a vast amount focuses upon the late eighteenth and nineteenth centuries, over one hundred years after Boyle and his contemporaries.[38] Second, of the studies to theorize the relationship between 'science' and imperialism, most are not intellectual histories and thus the *concept* of empire's scientific inheritance remains unexplored.[39] Chaplin, Drayton, Grove and others who do consider the role of natural philosophy in English empire-building in the seventeenth century have produced instructive scholarship. But the question remains. What role did natural philosophy play in shaping the very *concept* of empire?

Early Modern English Natural Philosophy

This book uses 'science' rather than natural philosophy in its title. This is not to imply that science and natural philosophy are identical but rather that natural philosophy shaped the origins of modern science. My task is to explore the origins of what is now popularly known as the relationship between 'science and empire', so I use the term 'science' to signify the book's contribution to these debates. The terminological quagmire that is the science/natural philosophy relationship often leads historians to use the term 'science' in the titles of their books. Here I follow the lead of historians like Michael Hunter and Joyce Chaplin who use the word 'science', yet are careful to historicize the nature of natural philosophy.[40]

Even the term 'natural philosophy', however, brings with it definitional problems. It is hard to identify exactly what constituted the subject in the early modern period. Nevertheless, I believe that we can plot a coherent tradition of natural philosophy in seventeenth-century England, centred upon the Baconian belief in the restoration of man's dominion over nature. In short, men like Bacon and Boyle were involved in a fundamentally religious project. In this sense, my research corroborates Andrew Cunningham's argument that seventeenth-century natural philosophy should be seen as 'a discipline and subject-area whose *role and point* was the study of God's creation and God's attributes'.[41]

The history of science has produced a wealth of literature on the idea of recovering Adam's dominion over the earth. Pioneered chiefly by Charles Webster, this project was part of the 'Puritanism and Science' thesis which argued that the rise of Puritanism was one of the engines of the Scientific Revolution in England. This thesis developed in response to two scholarly traditions. The first was the correlation between the Protestant work ethic and the 'spirit of capitalism' put forward firstly by Max Weber, and then by Robert Merton, who originally sought correlations between the rise of capitalist social classes and the practitioners of the new Science. [42] Merton's work was published in 1938. The second tradition emerged approximately two decades later in the 1960s and 1970s, and surrounded the claims made by Christopher Hill regarding the role of radical Puritanism in the English Revolution.

The foremost exponent of an argument influenced by Hill was Charles Webster, whose seminal work *The Great Instauration* was based upon his lectures to the Workers Educational Association in Leeds, 1968. Arguing that the origins of experimental natural philosophy consisted in the radical social reformist aims of a group of Puritan scientists, Webster reoriented the history of early modern English science, and did for this field what Christopher Hill did for the English Revolution. The substance of Webster's thesis was that Puritan intellectuals held the millenarian conviction of restoring Adam's prelapsarian dominion over the earth. 'It seemed', he wrote, 'that the fateful intellectual decline which had begun with the Fall of Adam might at last be reversed. The Puritan Revolution was therefore seen as a period of promise when God would allow science to become the means to bring about a new paradise on earth.'[43]

Webster pointed out the Biblical basis for the Puritan natural philosophers' millenarian hopes. It was not only Genesis 1:28 which captivated the Puritans, but also the book of Daniel, which stipulated that 'in the final days of the earth, knowledge will increase' (Daniel 12:4). The eschatological hope in the book of Daniel was one of the founding ideals of Francis Bacon's magnum opus, *The Great Instauration*. The inscription under the illustration of the Pillars of Hercules, on the title-page of the *New Organon*, was based upon the passage in Daniel: *Multi pertransibunt et augibitur scientia*, roughly translated as 'Many will travel to and fro, and knowledge will increase.' The opening up of the geographic world signalled the opening up of knowledge, such that the 'thorough exploration of the world (which so many long voyages have apparently achieved or are presently achieving) and the growth of the sciences would meet in the same age.'[44]

Natural philosophy assumed 'considerable significance in the Puritan programme and the Puritan intellectual became committed to a dedicated attempt to procure the return of man's dominion over nature'.[45] Webster, however, draws a neat historical line at the moment of the Restoration in 1660. With the institutionalization of natural philosophy that accompanied the emergence of the Royal Society, he assumes, the utopian element died out.

The Puritanism and Science thesis certainly has its critics. Lotte Mulligan argues that the distinct interest in science was held not only by Puritans but also by Anglican royalists. Furthermore, she argues that those characteristics which Webster termed 'Puritan' could equally be found among non-Puritans.[46] James and Margaret Jacob make the criticism that the dichotomy that Webster posited between Puritan science on the one hand, and the state-supported science of the Royal Society on the other, is unconvincing.[47] They argue instead that there existed a connection between the two, which manifested itself in a reformist dimension to the science pursued by members of the Royal Society. Boyle, together with John Wilkins, John Wallis, Christopher Wren and John Evelyn, 'continued to advocate and engage in the organized pursuit of experimental science. But they dissociated this project from any radical reform of the church, state, the economy or society'.[48] Rather, their Anglican reformism was manifested in a language of improvement, in order to inspire commerce and food production. As James Jacob puts it in a more recent article, 'the study of nature would make all men more pious by revealing God ... it would increase [economic] prosperity'.[49] Yet even though they recognize the significance of a programme of social and economic reform to the natural philosophy of the Royal Society, Webster and his critics do not place this intellectual tradition in the context of the history of England's imperial expansion and in the formation of 'empire' as a concept.

Puritanism was not the only intellectual context to encourage the pursuit of experimental natural philosophy. After all, the Royal Society was established in the Restoration, a period that produced its own set of conditions and problems for the pursuit of natural philosophy. As Michael Hunter has shown, the growth of a 'fashionable leisured culture focused on London'[50] was a significant enabling factor for the collective, organized and institutional pursuit of natural philosophy. The Restoration also placed several problems in relief, however. As Steven Shapin and Simon Schaffer demonstrated, one issue raised in particular by the Restoration Settlement was that surrounding the relationship between knowledge and social order.[51] This took the form of a general scepticism about the doctrinal nature of scholasticism which was intensified by the belief that scholastic philosophy, with its truth claims based upon the authority of texts, caused disagreement and civil unrest. Latitudinarians in particular found this repellent.

The issue became acute following the Restoration Settlement of 1660 which raised the question of how best to maintain public order and establish the legitimacy of political power which, as Shapin and Schaffer point out, was always entwined with the authority of knowledge.[52] One of experimental philosophy's responses to the problem was an empirical methodology for the production of facts – in a public space and with collective acclamation by other scientists. Barbara Shapiro and others have shown that the epistemological scepticism about the validity of scholasticism's claims generated the rise of an empiricist methodology in natural philosophy.[53]

The issue of establishing and verifying knowledge was peculiarly pronounced in an arena inaccessible to the majority of seventeenth-century natural philosophers. This theatre of knowledge was the New World. The problem of the reliability of knowledge is one of the most compelling in the scholarship.[54] It is not one of the central problems with which this book grapples, but my chapter on Boyle attempts to provide one modest point of entry into the debate. Boyle used his New World informants' connections to English colonies as a means of establishing their reliability.

Boyle's 'Outlandish Books' alert us to the excitement that the New World generated in natural philosophers. Recently, a number of historians of science have studied the importance of travel narratives to the early Royal Society. Daniel Carey approaches the topic by focusing on this issue of the reliability of knowledge. He shows that the narratives 'often masked rather than exposed problems of belief, testimony and evidence'.[55] Robert Iliffe places the Royal Society's interest in travel in the context of the growth of the Protestant state. In 'Foreign Bodies: Travel, Empire and the Early Royal Society of London',[56] Iliffe explores how, as the Protestant state grew more powerful, 'prejudice formed the basis of a dynamic process of "productive exclusion" [of foreigners] in which many practices which had once been part of a more pluralist outlook were deemed to be unacceptable'.[57]

Both Carey and Iliffe show originally and convincingly the significance of the narratives of travellers abroad to the scientific interests of the Royal Society. This line of enquiry could be enhanced, I would suggest, by extending our investigation into the relationship between the Royal Society's interest in the New World and the actual concept of empire that its fellows held. How did Royal Society fellows such as Boyle, Oldenburg and Edmond Halley articulate what they were doing? Did they, for example, believe they were contributing to a project of building an empire of knowledge?

Closely related to the Royal Society's endeavours to collect information and specimens from the New World was its creation of a repository (a museum) to store it all. This scientific imperial booty, as it were, is the other theme of my chapter on the Royal Society. I argue that, in its encyclopedic representation of the natural world, the Royal Society's repository became a tangible microcosm of an empire of knowledge.

The other dimension of scholarship on Adamic dominion concerns the significance of the Bible for seventeenth-century natural philosophers. Jim Bennett and Scott Mandlebrote's *The Garden, the Ark, the Tower, the Temple: Biblical Metaphors of Knowledge in Early Modern Europe*[58] is a compelling exploration of four Biblical metaphors and their influence upon early modern natural philosophy, particularly that of the Hartlib Circle. The metaphors of the Garden of Eden and the Ark, in particular, were influential in shaping natural philosophers' conception of Adam's empire. Bennett and Mandlebrote describe the belief that, before the Fall, the Garden of Eden's fruitful perfection meant that it enjoyed an ever-present springtime, and yet after the Fall the Garden's plants were made difficult to grow and its soil almost barren. After the Fall it was up to Adam's 'hard labour [to] make the Garden grow again, and restore the knowledge which humans had once employed to govern nature in paradise'.[59]

Despite the importance of the Bible during the early modern period, natural law is more frequently recognized as the major ideational source for legitimizing colonial expansion.[60] In a recent article, however, Peter Harrison demonstrates that seventeenth-century English colonists drew upon Biblical ideas to justify their colonial endeavours. These included the 'commission to preach the gospel to all nations',[61] and the Old Testament injunction to Adam, and then to Noah, 'Fill the earth and subdue it'.[62] In the seventeenth century this injunction was interpreted to emphasize the importance of agrarian labour. Having dominion over the earth 'is, by Culture and Husbandry, to Manure and make it fit to yield fruits and provision ... which is done by Planting, Earing, Sowing, and other works of Husbandry', wrote John White in 1656.[63] Harrison points out the connection between the Biblical warrant to subdue the earth and the development of formal doctrines of property and ownership. I am in complete agreement with Harrison's contention that 'there are clear parallels between rhetoric based on Old Testament narratives and John Locke's classic statement of the origins of private property in *Two Treatises of Government* (1690)'.[64] I aim to develop this idea by arguing that we can be quite specific about Locke's Old Testament inheritance: it was in fact a theory of *empire* as Adam's dominion over nature which helped underpin Locke's theory of property.

In his book *The Bible, Protestantism and the Rise of Natural Science*, Harrison turns his attention to the influence of Protestant hermeneutics upon the development of natural science. The Protestant practice of reading the Bible literally rather than emblematically generated an impulse to study nature.

The new conception of the order of nature was made possible by the collapse of the allegorical interpretation of texts, for the denial of the legitimacy of allegory is in essence a denial of the capacity of things to act as signs. The demise of allegory, in turn, was largely due to the efforts of Protestant reformers, who in their search for an unambiguous religious authority, insisted that the book of scripture be interpreted only in its literal historical sense.[65]

Harrison argues that literalism fostered the emergence of natural science in two ways: 'first by evacuating nature of its symbolic significance', and second by 'restricting the possible meanings of the biblical narratives of creation and the fall, in that they cannot be read as other than as enjoining upon the human race the necessity of re-establishing dominion over nature'.[66]

Harrison's point about the Protestant Biblical hermeneutics is one aspect of a larger shift in attitudes toward the natural world which characterized the late Renaissance. This shift is a good way to begin sketching the background to the relationship between natural philosophy and English colonization in the early modern period.

Natural Philosophy and English Colonization, *c.* 1550–*c.* 1700

The early modern period was characterized by fundamental shift in the perception of nature. As Paula Findlen puts it, 'before the sixteenth-century, nature was primarily a subject for allegory and commentary', yet during the sixteenth and seventeenth centuries, natural histories assumed a more recognizable form. Thenceforth, natural phenomena were not understood through their symbolic resonances but, rather, through systematic, taxonomical investigation.[67]

Admittedly, such a broad characterization is in some ways problematic. In Michel Foucault's *The Order of Things*, for example, this is an epistemic shift of grand importance, and yet hardly any explanation is given as to why it occurred.[68] It remains useful to note this shift, albeit cautiously, because it is one of the epistemological effects of what some scholars have termed the Scientific Revolution, a term I use here with the knowledge that it is problematic, but with the hope that its usefulness in giving us a name for a phenomenon outweighs these problems.

A second effect of the Scientific Revolution was that it produced a vision of Nature 'subject to rational laws which might be discovered and applied to human purposes'.[69] Nature was now subject to man's will to make it useful, an idea perhaps most famously expressed by Francis Bacon, whose aim was to use this knowledge as the foundation of a new and superior statecraft.

Salomon's House, on the island of Bensalem, in Bacon's posthumously published utopia, *New Atlantis*, was an official and well endowed institution in which scientists collectively pursued experimental philosophy in a public domain and for the public good.[70] Based upon Bacon's ideal for the advancement of learning, the institution would make possible the recovery of true knowledge. As Bacon wrote in *Valerius Terminus*, 'the true ends of knowledge ... [are] a restitution and reinvesting (in great part) of man to the sovereignty and power (for whensoever he shall be able to call the creatures by their true names he shall again command them) which he had in his first state of creation'.[71] It is hardly surprising that it is often thought that Salomon's House was a prototype for the Royal Society.

Founded in 1660, the Royal Society of London for the Promotion of Natural Knowledge emerged as the first official scientific institution in England. For several decades preceding the Restoration, however, a number of unofficial groups of like-minded gentlemen had met informally to discuss natural philosophy. One of the characteristics of the new experimental natural philosophy was, after all, the collaborative context in which it was practised. In 1645 a group of men including John Wallis and Robert Boyle met in London. Called the 'Invisible College' by Boyle in three letters, the group was the context for Boyle's introduction to natural philosophy.[72] By the end of that decade, many moved to Oxford, where, during the 1650s, the group centred upon Wadham College under the auspices of the college's warden John Wilkins.[73] By this time the group included Christopher Wren, Seth Ward and William Petty.

Although the exact nature of the connection is debated, many of these men went on to form the Royal Society, with Charles II as its patron, in 1660.[74] In 1662 the Society was granted a Royal Charter and a second one followed in 1663, which assured its permanency as an institution. During the late 1660s there was a plan to found a college for the Society, where its members could pursue adequately-funded research in either a laboratory or the observatory. The college, however, never came to fruition.[75] The Society did, however, form a number of committees directed to gleaning and managing information from far away places. In 1661, for example, Boyle, John Wilkins, Robert Moray, John Evelyn and Henry Oldenburg were all members of a committee formed to consider the questions to be enquired 'of the remotest parts of the world'.[76] In 1664 a Committee for Correspondence was formed, and its members were instructed to read various travel books.[77] When specimens were sent back to individual fellows, they were collected and kept in Gresham College where the Society met. Already possessing the title 'Curator of Experiments', Robert Hooke was appointed the Keeper of the Repository in 1663. Three years later, the Society purchased the cabinet of curiosities belonging to Robert Hubert, who claimed that he had travelled to the West Indies. In 1678 the botanist Nehemiah Grew was commissioned to produce a catalogue of the Society's repository. It was published three years later and was entitled *Musaeum Regalis Societatis*. Lack of funding and insufficient room in Gresham College for the store of curiosities, however, meant that the repository did not live up to expectations.[78]

What intellectual traditions inspired the Royal Society's collecting and its interest in natural history? A good point of entry into this question is the confrontation between the novelty of the New World and the ancient text of the Bible. As Peter Harrison has shown, seventeenth-century Protestantism developed a new textual hermeneutic in which the Bible was understood as literal history. But reading the Bible as history posed a number of questions. Where, geographically, was the Garden of Eden? Did it still exist? And whose descendants were the inhabitants of the New World? Thomas Burnet, in his *Telluris Theoria Sacra* (1681), argued that Eden could be found in the southern hemisphere, where the weather was perpetually temperate. Like many others, Burnet believed that before the Fall, the Garden of Eden enjoyed a constant springtime, and that seasons were imposed as a consequence of the Fall. The theme of the Fall from paradise and of its consequences for both mankind and nature was popular in the seventeenth century. The book of Genesis recorded God telling Adam that 'cursed is the ground for thy sake ... thorns and thistles shall it bring forth to thee' (Genesis 3:17–18). The poet Joseph Fletcher, in his collection of poems entitled *The Historie of the Perfect Cursed, Blessed Man* (1628), argued that those plants which were poisonous became so when the earth suffered the effects of man's fall. Similarly, the diarist John Evelyn believed that moss was a product of the Fall.

The fallen nature of the earth produced the redemptive possibility of agrarian labour. Through planting and husbandry, it was commonly believed, man could redeem himself by making the earth fruitful once more and reaping from it what John Locke called the 'conveniences for this life'.[79] Agricultural labour was a project of redemption: 'In the sweat of thy face thou shalt eat bread' (Genesis 3:19). Or, as John Evelyn explained in his *Elysium Britannicum*, man must 'recover ... by Arte and Industrie, which was before produced to them spontaneously; and to improve the Fruites of the Earth'.[80]

One of the forms that this labour took was the attempt to recreate the Garden of Eden itself. John Evelyn, in his *Kalendarium*, argued that man-made gardens should attempt to recreate Eden, 'as near as we can contrive them'.[81] Adam was believed to have been the first gardener, and he is represented as such in many seventeenth-century works, such as John Parkinson's *Theatrum Botannicum* (1640). As John Prest points out, the early botanic garden was conceived as a living encyclopedia; it was ordered and planted in rows, just as the original Eden was thought to have been. In fact, the Oxford Garden, which was opened in 1632, was square and its three acres were divided into four quarters representing the four corners of the earth.[82] By contrast, man now lived in a disordered world.

A fascinating illustration of one of the diverse thematic treatments of Adamic empire was the genre of handbooks of agricultural improvement. In 1649, Walter Blith's *The English Improver Improved, or the Survey of Husbandry Surveyed* appeared. Within a year it had been reprinted. Blith had republican connections. He was a captain in the parliamentary armies, and acted as an agent for the sequestration of royalist land. He also surveyed crown land.[83] But it was not only the republicans who used gardening as a tool of exerting dominion over the earth. Evelyn, a royalist, was asked by the Commissioners of the Navy to write an account of the cultivation of forest trees. *Sylva, or a Discourse of Forest-Trees, and the Propagation of Timber in His Majesties Dominions*, which was aided by various fellows of the Royal Society, was published in 1664.

The connection between husbandry and property extended beyond the creation of domestic gardens. As Patricia Seed has shown in her recent study, cultivating a garden was one of the most important ways that the English enacted their possession of the New World.[84] As the Massachusetts minister John Cotton wrote in 1630, 'in a vacant soyle, hee that taketh possession of it, and bestoweth *culture* and *husbandry* upon it, his Right it is'.[85] This dictum became famous primarily in the reception of John Locke's *Two Treatises of Government* but, decades before Locke published his work, English natural philosophers were well aware of the connection between husbandry – 'planting' – and legitimate property. The work that Richard Hakluyt presented to Queen Elizabeth in 1584, for example, was titled *Discourse on Western Planting*,[86] and it outlined various projects of colonial planting in the Americas.

Scholarship on the intellectual history of English colonization has paid excellent attention to the historical consequences of an agrarian theory of property. The idea that only cultivation constitutes ownership of land haunted the history of the English colonies from the seventeenth-century colonization of the Americas until the present day. In 1992 a momentous decision by the High Court of Australia in *Mabo and Others* v. *Queensland* found that an idea of native title (in the form of traditional ties and occupation of land) existed in common law. This overturned the theory of property which the British used to justify their dispossession of Aboriginal land since the colonization of New South Wales began in 1788.[87] One of the aims of this book is to show that the Lockean idea of property that legitimated this colonization was part of a larger theory of man's plenary dominion over nature.

Let us now move from the epistemic to the geographic, and outline the growth of England's colonization in the seventeenth century. English colonization was an outgrowth of the privateering expeditions of the late sixteenth century. The first expeditions were essentially licensed piracy; the aim was to wrest gold from the Spanish ships. English fishermen had for some time been fishing for cod in the icy waters of Newfoundland, and it was at St Johns in Newfoundland that Sir Humphrey Gilbert made the first claim of land for England in the New World in 1583. Three years later in 1586, Sir Walter Raleigh founded a colony on the island of Roanoke, in the marshes offshore from the modern-day North Carolina. A year later, however, the majority of the settlers had to be rescued by Sir Francis Drake. Drought, famine and attacks by the local indigenous people almost destroyed the colony. There was a second attempt to settle people on the island in 1587, but this colony also met its end soon after.[88]

Jacobean colonization was more successful. In 1606, James I granted a patent to colonize the east coast of North America between 34 and 45 degrees latitude.[89] The patents were won by a pair of joint stock companies, collectively known as the Virginia Company.[90] The first permanent settlement was in Jamestown in 1607; the colonization of New England followed soon after. The London Virginia Company gave a grant of land to Puritans which became the Plymouth Colony, founded by those from the *Mayflower* in 1620. Maine became a province in 1622, New Hampshire was settled the following year, Massachusetts Bay in 1626, and Connecticut, which was settled by Puritans from Massachusetts, in 1633. Formal government was established on the island of Acquidneck, which was renamed Rhode Island, in 1644.

England was also active in the Caribbean, and it would be a mistake to assume that the North American colonies of settlement were more important than the

Caribbean colonies of trade. It was the Caribbean that offered the abundance of raw materials that England needed. The sugar plantations in Barbados made it the most prosperous English colony by far. The Virginia Company laid claim to Bermuda in 1612, after an initial English settlement there in 1609. Settlements were then established at St Kitts in 1623–4, Barbados in 1627–8, Nevis the following year, Barbuda and Antigua in 1632 and Anguilla in 1650. Five years later Oliver Cromwell's troops under the commandership of William Penn Sr wrested Jamaica from the Spanish.

The administration of the colonies provided much cause for discontent. Between 1650 and 1657 there were several Committees of Trade, and these were replaced in 1660 by the newly established Council for Trade, and the Council for Plantations. It was the councils' role to apprise themselves of the condition of the colonies in terms of their administration and needs, and also to instruct colonial governors on issues of law and welfare.[91] In 1672 these two bodies were reconstituted as the Council of Trade and Foreign Plantations. John Locke was appointed the Council's Secretary in 1673–4. Locke was also the most influential member of the subsequent body, the Board of Trade, which was established in 1696.

In 1651, Cromwell's Parliament introduced the first of a number of Navigation Acts which were designed to restrict foreign shipping and ensure that the raw materials from England's colonies were traded for her own benefit. This would place pressure on the Dutch who had dominated the carrying trade.[92] The Acts enshrined the view of the colonies as a source of raw materials for England, and they were received with a good degree of resentment. The Puritan population of Massachusetts Bay, who had been trading with the Dutch of their own accord, ignored the Acts, while Barbados petitioned London to be excused from them. The Acts were only loosely enforced until the mid-1670s. From then until the Glorious Revolution, however, the absolutist tendencies of Charles II and then James II increased the control of England over her American dominions, particularly in regard to enforcing the Navigation Acts. In 1686, James II declared the Dominion of New England, which brought the colonies of New England under the one governor, Edmund Andros. The Dominion was repealed after the Glorious Revolution.

As the historiography of early American history has charted, the situations in New England, the Chesapeake and the Caribbean islands markedly differed from one another. The Chesapeake struggled with disease, conflict with Algonquian Indians and famine. The Virginia Company collapsed and had its charter revoked in 1624, and Virginia became a crown colony. The colonies in Puritan New England, by contrast, were quite successful. They traded independently with the Dutch and their population grew. By 1640 over 20,000 people had settled in Massachusetts.[93] In contrast to the prosperity of New England, the island of Montserrat acted as a penal settlement where Irish Catholics were forcibly moved in 1631s. More dissidents were transferred after the battle of Drogheda in 1649. This human trafficking is perhaps the most graphic illustration of the way in which English colonization entwined the histories of peoples and countries across the Atlantic and Caribbean.

Historical Change

The influence of natural philosophy upon the intellectual origins of the British Empire developed through three phases, from the late sixteenth to the early eighteenth century. When Francis Bacon put the Pillars of Hercules on the frontispiece of the *New Organon* in 1620, his conception of the New World was as a storehouse of information which could be explored by travellers and merchants. Thomas Harriot and Richard Hakluyt held the same view. For this generation, the knowledge offered by the New World was quite a separate issue from that of colonization. Indeed, for Bacon, colonies posed the dangerous moral threat of corruption.

By the time members of the Hartlib Circle were agitating and writing in the 1630s and 1640s, however, the English colonies were becoming viable enterprises. Their inhabitants began to survive, reproduce and establish towns. The population in New England multiplied. This was the historical moment in which colonies began to impress themselves on the consciousness of natural philosophers, through the expeditions of men such as John Tradescant the younger (*c.* 1608–62), whose collections formed the basis of what became the Ashmolean Museum in Oxford. Tradescant travelled to Virginia three times, first in 1637, then in 1642 and finally during 1653–4. Moreover, a number of gentlemen interested in science, such as John Winthrop Jr, the Governor of Connecticut, actually lived in the colonies. The success of the colonies as viable places to live established the material conditions that made possible the communication between the Atlantic periphery and natural philosophers in the metropolis.

For all the Hartlib Circle's hitherto overlooked fascination with the Atlantic world, their interest did not extend to conceptualizing any systematic relationship between an empire of knowledge and the Atlantic colonies. It was Robert Boyle who, in the Restoration, first put forward a programmatic suggestion for the way in which man's dominion over nature could be restored by fostering a relationship between naturalists and the English colonies. This relationship would enable England to harness a wealth of information from the New World that would improve natural philosophy. During this historical moment a new idea emerged that characterized the second phase of development. The colony could serve as a space for the production of natural knowledge.

Boyle's ideas regarding the beneficial coexistence of colonies and natural philosophy were put into practice by the Royal Society. The Society's members made

constant requests to be sent information about natural history from their correspondents in the New World. Not only were letters containing this information
sent back from American colonies such as Massachusetts Bay and Virginia, but
some men, such as Thomas Hill, shipped back to London wooden boxes containing plant and animal specimens. Other natural philosophers such as Sir Hans
Sloane and the astronomer Edmond Halley travelled to the Americas themselves
and collected specimens. Sloane travelled to Jamaica and St Helena; he returned
with over 800 plants which eventually formed the basis of the British Museum.
The Atlantic transportation of wooden boxes of natural artefacts and letters
about plants was animated by the ideal of recovering an empire of knowledge.

This kind of practice illuminates one of the most troubling problems with
which this book grapples: the tension between the universal language of Adamic
empire, which, of course, belonged to man, and the very *English* empire of
knowledge which the natural philosophers intended to create. They maintained
the belief that England adopted the Protestant mantle for all mankind. Moreover, the way that Adamic ideas were put into practice benefited the English.

This is not to say that the natural philosophers were experiencing any imperial 'absence of mind', to quote John Seeley. After all, Boyle knew exactly what he
was doing when stipulating a means through which the English colonies could
aid the philosophers' empire of knowledge. Rather, this disjunction between
intention and outcome illustrates the way that the processes of empire-building
encompassed a European intellectual tradition that was not specifically British.

The third and final development was the sophisticated incorporation of the
epistemic idea of empire into a philosophy which would influence the ideology
and practices of the British Empire for centuries. This was John Locke's refashion-
ing of epistemic empire into a theory of 'improvement'. When we explore Locke's
philosophy we find it imbued with the sense of recovering the fruitfulness of the
fallen earth for the benefit and enjoyment of mankind. For Locke, the purpose
of natural philosophical knowledge is to endow man with the faculties to 'apply
[parts of the universe] to our uses and make them subservient to the conveniences
of our life, as proper to fill our hearts and mouths with praises of [God's] bounty'.[94]
This is the idea of epistemic empire; it is the injunction to fulfil God's plan by
improving the earth and therefore returning man to his intended empire over it.

Locke put his ideal of improvement into practice. He worked as a secretary to
the Lords Proprietors of Carolina from 1668, served as Secretary to the Council
for Trade and Plantations and was subsequently the most influential member of
that Council's successor, the Board of Trade. The epistemic idea of man's empire
over nature animated the ideological origins of the British Empire. We still feel
its reverberations today.

1 'IN A PURE SOIL': FRANCIS BACON'S EMPIRE OF KNOWLEDGE

> The end of our Foundation is the knowledge of Causes, and secret motions of things; and the enlarging of the bounds of Human Empire, to the effecting of all things possible.
>
> Francis Bacon, *New Atlantis* (1627)[1]

The aim of Francis Bacon's utopian society Bensalem is generally accepted as emblematic of his natural philosophy. The recovery of 'Human Empire' was an idea to which Bacon referred many times across the corpus of his work, from the 'The Masculine Birth of Time' (1601–2) to the *New Atlantis* (1627). Despite its significance, scholars have not placed Bacon's ideal of 'Human Empire' in the context of England's Atlantic colonial ventures. Yet Bacon had much to say about English colonization and exploration of the New World. He was a member of both the Virginia Company and the Newfoundland Company; he wrote essays on plantations and empire; he argued for the general naturalization of Irish subjects; and he gave advice to the Earl of Essex, Queen Elizabeth and then James I on the administration of the Irish plantations. Most significantly, Bacon used the term 'empire' to describe the central tenet of his project, *The Great Instauration*. This was the ideal of restoring man's original dominion over nature.

Was there any relationship between Bacon's conception of man's prelapsarian empire and his interest in the New World and colonization? The answer is both surprising and complex. Bacon's vision of the restoration of man's dominion over nature contained an important role for knowledge gleaned in the New World but, importantly, this knowledge was to be collected through exploration rather than through establishing colonies. There is no connection in Bacon's work between colonies and the restoration of man's empire, the latter of which is an epistemological rather than territorial pursuit.

Why, then, begin this book with a chapter on Bacon? Precisely because the absence of connection between colonies and man's prelapsarian empire in Bacon's work enables us to understand the degree and nature of historical change over the seventeenth century. In the work of Bacon's successors – the Hartlib Circle, Robert Boyle, the Royal Society and John Locke – the restoration of man's Adamic empire became increasingly tied to the pursuit of English colonization. This is because over the course of the seventeenth century, one aspect of Adamic empire – the recovery of the earth's fruitfulness – became more important than the recovery of man's natural knowledge, and this agrarian dimension was intimately connected with the colonial 'improvement' of land. For Bacon, however, Adamic empire is about the recovery of perfect natural knowledge. Exploring Bacon's work makes it possible to chart the historical shift in the conception of man's dominion over the course of the seventeenth century.

Bacon's work on 'Humane Empire' and its recovery is also illuminating on its own terms. The New World contained knowledge that would greatly advance the pursuit of restoring man's natural knowledge, but Bacon doubted whether colonization was the best method for recreating this epistemic dominion. In fact, Bacon held humanist anxieties about the morality of colonization, and he proposed an explicitly non-colonial ideal of collecting knowledge in the New World. This is idealized in Bacon's utopia, the *New Atlantis* (1627). Contrary to a great proportion of the scholarship which views this text as legitimating colonial possession, I will argue that Bacon's *New Atlantis* challenged the idea that colonization was the best means of utilizing the New World in order to advance knowledge. Rather than establishing settler colonies, the people of Bensalem collected knowledge through the exploratory travels of scientists. These voyages of discovery represented Bacon's ideal of a non-colonial means of recovering an epistemic empire, and they mirrored his real-life hope that the explorations of Elizabethan explorers could yield useful scientific information.

Bacon's Language of Empire

It is useful to begin by outlining briefly the way that Bacon used the term 'empire' and noting some important linguistic ambiguities. Bacon's primary use of the term was to denote man's original power over the earth. The full title of Bacon's 'The Masculine Birth of Time' (1602–3) is a good example. The Latin reads 'Temporis Partus Masculus Sive Instauratio Magna Imperii Humani in Universum',[2] which can be translated as 'The Masculine Birth of Time or, The Great Instauration of the Empire of Man Over the Universe'. The full title of 'Temporis Partus Masculus' illustrates Bacon's primary use of the term empire to mean the renewal (*instauratio*) of man's original dominion over nature.

Four important points need to be made about Bacon's use of the term 'empire'. The first is that Bacon did not use the word to denote England's overseas territories. In the early to mid seventeenth century, the meaning of 'empire' as territorial dominion of course existed, deriving from the Roman tradition in which 'empire' referred to Rome and her colonies. The key point here is the difference between 'empire' as a straightforward concrete noun, and empire as an abstract noun. At the time Bacon wrote, empire denoting a set of overseas colonies was used only to refer to the Roman Empire. It was not used to refer to England's patchwork of Atlantic colonies until much later in the seventeenth century.[3]

Empire as an *abstract* noun, however, in the sense of having an empire over something, was more commonplace and *could* be used to denote power over the three kingdoms. This is the second point: not every time Bacon uses the term empire was he referring specifically to the plenary dominion of mankind over the earth. There are instances, for example, in which Bacon used the term as a straightforward metaphor for uncurtailed sovereignty, that is, in the Henrician tradition of 'this realm of England is an empire'.[4] The difference between the empire of Adam, and 'this realm of England', however, is one of context rather than meaning. That is, empire still denotes unlimited power. The use of the term to denote Adam's empire, then, is a reference to a specific instance of that power as it related to man's original dominion of knowledge over the creatures and resulting power over the earth. The fact that empire is a metaphor for unimpinged power draws our attention to a third point. For Bacon, dominion and empire were in most contexts synonyms.

The fourth point which we must keep in mind is that there is a profound ambiguity in Bacon's work concerning who would reclaim man's empire over the earth: mankind or England? Bacon's language when he refers to Adam's dominion, as we will see, is universal. Adam is a synecdoche for mankind. We know, however, that the overwhelming intention of Bacon's work was to advance learning in order to enhance the power of England rather than mankind. This is an unresolved tension in Bacon's work.

Although Bacon's use of the language of empire was varied, the task of understanding his conception of Adamic empire is aided by the fact that he was not doing anything lexically new. What *was* new about Bacon's work was his vision for the means of restoring man's original empire through the collaboration of natural philosophers sponsored by the sovereign. The fact that this vision was accompanied by a scepticism about colonization compounds the significance of Bacon's work to our understanding of the development of the concept of empire.

Bacon Scholarship Past and Present

Francis Bacon is not usually a figure included in histories of the British Empire. The lack of dialogue between the history of science and the intellectual history of the British Empire in the sixteenth and seventeenth centuries is for the most part responsible. In recent years, however, a small number of histories of English colonization have included Bacon. One such is Andrew Fitzmaurice's intellectual history, which brings the hitherto neglected context of humanism to bear upon colonization from the Tudor period onwards. Contrary to the assumption that English colonization was driven by the desire for profit, Fitzmaurice shows that early colonizers of America were often 'skeptical of profit and nervous of foreign possessions at the same time that they saw both as possible sources of glory'.[5] Fitzmaurice identifies a civic humanist language as the basis for Bacon's concerns about 'displanting' indigenous people in his essay 'Of Plantations'.[6]

In David Armitage's *Ideological Origins of the British Empire*, the main significance of Bacon is as an author writing about the relationship between the three kingdoms in the mid sixteenth to early seventeenth century. Armitage shows Bacon 'echoing King James' aspirations' in seeing the Ulster plantation as 'a second brother to Union',[7] and then comparing the Virginia colony unfavourably to the Ulster plantation.[8]

While Bacon has not made much of an appearance in histories of British colonization and empire, he has a copious body of secondary literature devoted to his philosophy. The vast majority of this scholarship, however, does not see ventures into the Atlantic as playing any significant role in Bacon's oeuvre. Aside from a few studies of the *New Atlantis* which see it, wrongly in my view, as a tract legitimating colonization, the common position is a fairly vague assumption about the analogous opening up of intellectual and geographic worlds.

Bacon scholarship has, however, paid close attention to the philosopher's ideal of restoring dominion over the world. This was the aim of Bacon's project, the *Instauratio Magna*, into which he incorporated most of his natural philosophy. The idea of the Great Instauration is the basis for Charles Webster's seminal study of the Puritan influence upon seventeenth-century English science. For Webster, Bacon provides the original articulation of a project which shaped the following century of the natural philosophical innovation. 'The fragmentary philosophical system bequeathed by Bacon became for puritan intellectuals both the basis for their conception of philosophical progress and the framework for their utopian social planning.'[9] In short, the Great Instauration was the return of man's complete dominion over nature.[10]

Webster notes that quotations from the book of Daniel appear frequently in Bacon's early writings.[11] Daniel 12:4 is the source of Bacon's epigraph on the title page of *New Atlantis*, 'multi pertransibunt et augebitur scientia' – many shall run to and fro and knowledge shall be increased. Webster is one of the few historians to recognize Bacon's interest in the travels of his contemporaries Thomas Hariot and Walter Ralegh. This was particularly evident in *New Atlantis*, in which Bacon was 'undoubtedly influenced by the imaginative and optimistic accounts of America and the Islands of the West Indies, published by Hakluyt, Raleigh and Harriot, or even by the stream of propaganda on the wonders of the New World issued by the promoters of the Virginia Company between 1606 and 1624'.[12] The focus of Webster's book, however, is not upon Bacon but upon his legacy for the scientific revolution.

The work in which Bacon first articulated his idea of restoring dominion over the world was his 'Temporis Partus Masculus' ('The Masculine Birth of Time'), written between 1601 and 1602, and first translated into English by Benjamin Farrington. The work is a critique of Aristotelian philosophy which relies heavily upon Biblical symbolism.[13] Farrington's commentary explores the idea, probably derived from Giordano Bruno, that 'a new order of events was at hand, the specific quality of which would consist no longer in a mere *imitation* of nature but in her *domination* by man'.[14]

Bacon's belief in the possibility of dominating – rather than imitating – nature was based upon an epistemological premise. This was the connection between man's ability to know, and his capacity for making, for making was a way of dominating nature by reproducing its effects. Here the contrast between Bacon's new method and scholastic philosophy is obvious. As Antonio Perez-Ramos has shown, Bacon's philosophy is part of a tradition of 'Maker's Knowledge'.[15] Brian Vickers explains that Bacon's attitude towards rendering philosophy productive of human action 'undoubtedly reflects that tendency in Renaissance thought to emphasize the power of the human will to make the *vita activa* the dominant model for man as a social animal'.[16] Thus 'to know something (a natural phenomenon) amounts to being able to (re)produce that very phenomenon on any material substratum susceptible of manifesting it'.[17]

The roots of Bacon's idea of instauration are also informative. Charles Whitney demonstrates that the word instauration 'alludes to the Vulgate's [Bible's] use of the term, which denoted the restoration of Solomon's Temple'.[18] The idea of instauration was not new to Bacon. Whitney showed that he 'may have adopted the term after seeing Tycho Brahe's *Progymnasmatum Instauratae Astronomiae* (*Exercise of Renovated Astronomy*, 1602), which, like the *Instauratio Magna*, is dedicated to King James'.[19] Whitney argues that Bacon overlays a Roman idea of the cyclical renewal with a progressive and linear understanding of time. The Roman emperors used the term instauration as a way of claiming they would renew everything great about their empire.[20] It thus had a cyclical meaning. In the Biblical texts, however, there is a providential and linear conception of time, in which the renovation of Solomon's temple by the post-exilic Hebrew community signifies not only the rebuilding of Israel, but also a 'Christian instauration of all things in the apocalypse'.[21] Whitney argues that Bacon brings together these two conceptions of instauration. His programme 'aims not just to complete the instauration of learning hailed by Du Bellay, Erasmus, and Rabelais, but also to launch a period of growth ending only with apocalypse'.[22]

In addition to the Biblical and classical contexts for the idea of instauration, there is a third useful interpretive context, that of Renaissance alchemy and the occult. The occult origins of Bacon's idea of dominion have been explored by Paolo Rossi in *From Magic to Science*. Rossi reveals the influence of Renaissance magic and alchemy in shaping Bacon's idea of man as nature's 'servant and inter-preter'.[23] Cornelius Agrippa, in a text that Bacon certainly read, defined natural magic as the power to manipulate nature. This power, however, was based not upon incantations, but upon an understanding of the workings of nature. Natural magic, therefore, 'is that which having contemplated the virtues of all natural and celestial things and carefully studied their order proceeds to make known the hidden and secret powers of nature ... For this reason magicians are like careful explorers of nature only directing what nature has formerly prepared ... so that these things are popularly held to be miracles when they are really no more than anticipations of natural operations'.[24] As Rossi shows, 'this definition of natural magic is akin to Bacon's concept of an art following faithfully in the footsteps of nature, incapable of miracles because it is a human art with human limitations'.[25]

A fourth interpretative context for Bacon is that of Renaissance humanism. Stephen Gaukroger has shown persuasively that Bacon's project for the reform of natural philosophy was founded upon a new role for public knowledge.[26] The philosopher was transformed 'from someone whose primary concern is with how to live morally into someone whose primary concern is with the understanding of and reshaping of natural processes'.[27] Likewise, Bacon made 'the first systematic, comprehensive attempt to transform the epistemological activity of the philosopher from something essentially individual to something essentially communal'.[28] The aim of this philosophy was the creation of a powerful, properly administered state.[29] In this sense, Gaukroger is developing a line of enquiry advanced by Julian Martin.[30] Gaukroger also argues that 'there was something especially appropriate about Bacon's outlook for the colonizers of the New World',[31] and that Bacon viewed public knowledge 'as being something which might serve the monarch, in some ways on a par with territorial conquest'.[32] If we add to this line of analysis an exploration of Bacon's idea of empire, we arrive at a more complex set of observations. We discover, for example, Bacon's series of anxieties about colonization, and a disjunction between his conception of empire and his views on colonization.

Markku Peltonen has also placed Bacon in the context of his humanism and classical republicanism. Arguing that the corpus of Bacon's work should not be seen as constituting a single coherent project, Peltonen suggests that Bacon's classical republican moral thought indicates that we should read him primarily as a rhetorician who frequently argued for contrary ideas.[33] This is the reason why there were often contradictions in his work, and why we need to see Bacon's political philosophy and his scientific programme as constituting two separate projects. Peltonen's article is a critique of an earlier strand of scholarship which read both Bacon's science and his politics as part of one 'imperial' project. Although this is not a point made by Peltonen, I would suggest that those scholars who make the 'imperial' argument about Bacon do so by employing two wholly anachronistic assumptions: first, that there existed some reified 'ism' called 'imperialism' and, second and more importantly, that at this historical moment, England's colonial possessions overseas were called an 'empire'. For Howard B. White, for example, there existed a maritime and commercial imperialism which Bacon advocated.[34] In the late sixteenth and early seventeenth centuries, however, there was no conception of a colonial, expansionist – let alone maritime and commercial – 'imperialism'. The concept of empire is vital to understanding Bacon's work, but not in terms of a territorially expansionist 'imperialism', but rather to denote the renewal of man's plenary empire over nature.

Peltonen mounts a convincing critique of the strand of Bacon scholarship which misconstrues his work as constituting one great project of imperial, territorial expansion.[35] He is also astute in pointing out Bacon's classical republicanism in the context of his political philosophy, and the fact that this often led to contradictions between his politics and his natural philosophy. We can, however, note one important consistency between these two fields of Bacon's thought. Bacon's natural philosophical project to foster the reconstruction of man's empire of knowledge was entirely consistent with his humanist position on colonization. The thesis of this chapter is that Bacon's classical humanist anxieties about colonization were the reason why he envisaged a specifically non-colonial means of restoring man's empire of knowledge. Travel narratives, Bacon believed, were the means through which the necessary information could be gleaned from the New World.

Paula Findlen has contributed a great deal to our understanding of the development of natural history and collecting in early modern Europe. One of her arguments about Bacon, however, is slightly anachronistic. 'Like many prominent Elizabethans [Bacon] found the image of an English empire to be particularly appealing; economic expansion, political might and mastery of nature all came together in the possession of the New World.'[36] Findlen's statement assumes an anachronistic concept of empire to refer to England's colonies in the New World. As I pointed out earlier, the use of the term 'empire' to denote England's overseas colonies did not occur in Bacon's time and, as far as we know, probably not until the late seventeenth century. It is incorrect to assume that Bacon saw a connection between epistemic empire and a territorial empire of New World colonies. A degree of caution is needed in understanding the complex, and often ambiguous, development of the meaning of empire. The aim of this chapter is to show that in Bacon's work there is no necessary connection between man's empire over nature and that of England's colonies. In fact Bacon envisioned a specifically non-colonial means of restoring man's empire in the New World through explorative travel.

Background and Context

The best way to introduce Bacon's philosophy and colonial interests is with his own words. Bacon defined his natural philosophy as 'the Inquiry of Causes and Production of Effects'.[37] That is, natural philosophy must be of use for the improvement of man's life on earth. Bacon's emphasis upon the aims of knowledge is the first of three important characteristics of his philosophy.

As Perez-Ramos has shown, Bacon's idea that natural philosophy should produce 'works' was epistemologically new; man's knowledge is defined in terms of his ability to reproduce what he studies.[38] What distinguished Bacon's *New Organon* from the Aristotelian 'old' Organon was the idea, as Bacon expressed it in *Cogitata et Visa*, that 'it is by witness of works, rather than by logic ... that truth is revealed and established ... whence it follows that the improvement of man's mind and the improvement of his lot are one and the same things'.[39]

The ability of natural philosophy to be productive of useful works presupposed that knowledge could be advanced over time. One of the means through which this advancement would occur was the construction of meticulous and empirical natural histories. A major weakness in the existing state of knowledge, Bacon believed, was the unsystematic and unprofessional pursuit of natural history in Tudor England. Collecting natural curiosities was a popular pastime of the aristocracy, but Bacon viewed such collections as focusing only upon the curiosity of the artefact; these natural histories were aimless, their objects were decontextualized, and the trivial collections did not advance knowledge of their subject. As Paula Findlen points out, they symbolized the aesthetic, feminized court culture of the Elizabethans.[40]

Most of Bacon's writing in the field of natural history occurred during the brief period in the 1620s before his death in 1626. In 1622 he began writing six natural histories that would function as prototypes of quantitative empirical natural philosophy. In only three of these works did Bacon get further than the preface: The *Historia Ventorum*, which was published that year; the *Historia Vitae et Mortis*, published the following year, and the *Historia Densi et Rari*, which was published posthumously in 1658. The *Historia Vitae et Mortis* was particularly quantitative, and its statistical analysis of life and death foreshadowed William Petty's political arithmetic in Ireland. Bacon's influence upon the writing of quantitative and empirical natural histories was not properly evident until the emergence of the Royal Society in the Restoration.

A second distinctive characteristic of Bacon's natural philosophy was his emphasis upon the importance of public knowledge. Natural knowledge should be public both in its production and also in its use for the good of the state. In fact, Bacon's political career can be thought of as a series of attempts to share his knowledge with the monarch. He had some trouble gaining Elizabeth's attention even though he wrote his first piece of advice to her as early as 1584 when he was only twenty-three. Under Elizabeth, Bacon served in the House of Commons from 1581, but he did not advance further during her reign. Under James I, however, Bacon's fortunes rose. Upon James's accession to the English throne in 1603, Bacon was knighted. Five years later he became Solicitor-General. During James's reign, Bacon became interested in the Virginia Company and its fledgling colony in the Chesapeake, which was first settled in 1607. From 1609, the archives of the Virginia Company list Bacon as a shareholder, and he also bought shares in the Newfoundland Company. Together with the Earl of Pembroke and the Earl of Southampton, Bacon was a council member of the Virginia Company, and in fact Bacon's name appears on the Second Charter of Virginia, dated 23 May 1609.[41] Bacon's membership of the Virginia Company and the Newfoundland Company is something rarely commented upon.

Bacon's career continued to shine under James. He became Attorney-General in 1613, Privy Councillor in 1616, Lord Keeper in 1617 and Lord Chancellor the following year. Here Bacon served James until he was impeached supposedly for judicial corruption. In reality, Bacon was a scapegoat for the Duke of Buckingham. He was stripped of his offices; it was an ignominious downfall. Bacon's fall from grace, however, sparked the most productive years of his life. Although the first two parts of the *Instauratio Magna* appeared in October 1620, just before his downfall the following winter and spring, the rest was published over the next six years. The work remained incomplete upon his death, and probably could never have been finished by Bacon even had he lived another decade, but the *Instauratio Magna* was nonetheless a colossal project. Divided into six parts, the work brought together Bacon's interest in natural history with his overarching concerns for the twin aims of restoring mankind's original empire, and using this knowledge to establish England as a powerful, centralized state.

The first part was a survey of existing knowledge, together with a plan for its reform. For this purpose Bacon revised and translated into Latin his 1605 work, *The Advancement of Learning*, which became *De Augmentis Scientiarum*. The second part was the *Novum Organon*, his new method for natural philosophy. The third part aimed to be a comprehensive compilation of 'the *Phenomena of the Universe*'.[42] This meant a collection of natural data in the form of natural histories. It was in this section that Bacon attempted to write his own natural histories, six of them, that would serve as prototypes.

The fourth part was designed to put into practice his methodology. This section, however, was left almost untouched. Bacon in fact admitted in the preface to this section, the *Abecedarium Novum Naturae*, that the data he possessed were insufficient for the section's completion. Parts five and six were also left primarily unwritten. Part five was intended to list discoveries that Bacon himself had made, and part six, the completion of which, Bacon admitted, was 'beyond my power and expectations', was the final summation of his project.[43]

The third, and perhaps most fascinating, characteristic of Bacon's philosophy was his consciousness of the historical context of his own work. From early in his philosophical career, Bacon presented his work as a product of the era in which he wrote, describing it as the masculine child of time, in the 'Temporis Partus Masculus' (1601–2). The sense that the sixteenth century was radically different from the ages that preceded it was common among Elizabethan writers and particularly adventurers. Nothing brought home the idea more than the voyages to the New World. Bacon made frequent references to the discovery of the Americas. The eighty-fourth aphorism in the first book of the *New Organon* compels his audience to take into account the fact that 'by prolonged voyages and journeys (which have become prevalent in our times) many things in nature have been disclosed and found out which could shed new light on philosophy'.[44] The twin metaphors of geographic and epistemic exploration occur frequently in Bacon's work, as do Biblical allusions to Adam and the rebuilding of Solomon's Temple. This was an especially pertinent and politically useful allusion given James I's identification with the Israelite king.

These ideas, however, were not unique to Bacon. In fact, metaphors like the reconstruction of Solomon's Temple were popular during the sixteenth century.[45] As David Sacks has shown, Richard Hakluyt imagined the symbolic re-establishment of the Temple through the processes of geographic discovery.[46] Like Bacon, Hakluyt envisaged an apocalyptic process that would produce 'a healing of the world through the emulative recovery of knowledge and of the command over nature that had been lost by Adam at the Fall'.[47]

The novelty of Bacon's philosophy, therefore, was not his vision of the restoration of Adam's empire of knowledge. This was shared by men like Hakluyt, Thomas Harriot and John Dee, among others. Rather, Bacon's novelty was his thorough vision for its achievement. His work is bold, ambitious and original. For what it lacks in empirical basis and in completion, it compensated for in its aspiration.

Bacon's Concept of Empire

Bacon's idea of restoring mankind's dominion over the earth was a concept of empire. While scholars have paid close attention to Bacon's aim of recovering man's lost power over the Creation, they have largely done so without examining Bacon's language closely. In this section I outline Bacon's use of the term 'empire', which is for the most part overlooked by the scholarship because of the way in which the Latin term 'imperium' was translated into non-imperial terms in English. Let us begin by introducing Bacon's idea of human empire that he derived from an understanding of Adam's original omniscience. In the *New Organon*, Bacon argues that the purpose of natural philosophy is to improve man's estate. This is most memorably articulated in his third aphorism: 'Human knowledge and power come to the same thing, for ignorance of the cause puts the effect beyond reach. For nature is not conquered save by obeying it; and that which in thought is equivalent to a cause, is in operation equivalent to a rule.'[48] Bacon referred to this idea frequently, and it is closely tied to man's loss of his original empire at the moment of the Fall from Eden.[49] Bacon gives an elegant and powerful explanation:

> For by his fall man lost both his state of innocence and his command over created things. However, both of these losses can to some extent be made good even in this life, the former by religion and faith, the latter by the arts and sciences. For the curse did not quite put creation into a state of unremitting rebellion, but by virtue of that injunction *In the sweat of thy face shalt thou eat bread*, it is now by various labours (not for sure by disputations and the idle ceremonies of magic) at length and to some degree mitigated to allow man his bread or, in other words, for the use of human life.[50]

Bold though it was, Bacon was anxious to ensure that his natural philosophical project was not irreligious or heretical. He was careful to explain that it was not the search for *natural* knowledge that precipitated the Fall; it was the temptation by *moral* knowledge. As Peter Harrison has shown, Bacon 'modified an existing view of the moral legitimacy of knowledge of nature in order to provide rhetorical justification for his proposed instauration of learning:'[51]

> For it was not that pure and unstained knowledge of nature, the knowledge by which *Adam* gave names to things according to their kind that prompted or occasioned the Fall, but that ambitious and importunate craving for moral knowledge to judge of good and evil so that man might revolt from God and give laws to himself was the ground and measure of temptation.[52]

Bacon made similar comments in *Filium Labyrinth*, where he ensured that 'all knowledge and specially that of natural philosophy tendeth highly to the magnifying of the glory of God in his power, providence, and benefits.'[53] Natural philosophy would reveal the power – rather than the will – of God in nature, so this was legitimate knowledge. Indeed, as Bacon pointed out several times in his work, God in fact encouraged a kind of hide-and-seek game with the great discoveries of the earth. '*The glorie of God is to conceale a thing, But the glorie of the King is to find it out* as if according to the innocent play of Children the diuine Maiestie tooke delight to hide his workes, to the end to have them found out.'[54] Bacon cites this phrase several times in his work, and this use of the same phrases and terms alerts us to the coherence of Bacon's ideal of Adam's epistemic empire.

Let us explore Bacon's use of the term 'empire' to describe his project of restoring man's original dominion over the creatures. Bacon frequently used empire – *imperium* – to describe his project. The scholarship has overlooked this, however, and this is chiefly due to the fact that Bacon's term *imperium* is translated into English as 'dominion'. It is true that the English terms 'empire' and 'dominion' were often used as synonyms in the seventeenth century. I do not want to suggest, therefore, that there has been an egregious error in misinterpreting Bacon's meaning. I do think, however, that had Bacon's term *imperium* been translated as 'empire' rather than 'dominion', scholars might have been alerted to the prevalence and significance of the idea of empire in Bacon's work. I would suggest that the reason the scholarship has not, for the most part, investigated Bacon's idea of empire is that when contemporary scholars see the term 'dominion' we are apt *not* to automatically make the theoretical connection to 'empire'.

Let us consider this translation issue in one of Bacon's earliest works, the 'Temporis Partus Masculus' ('The Masculine Birth of Time', 1602–3). This work was not translated into English in the nineteenth-century edition of Bacon's work compiled by James Spedding, Robert Ellis and Douglas Heath. 'Temporis Partus Masculus' was translated into English for modern readers by Benjamin Farrington in 1964. The text was included in Farrington's *The Philosophy of Francis Bacon: An Essay on Its Development From 1603–1609 with New Translations of Fundamental Texts*. The title-page of Bacon's 'Temporis', which includes a subtitle, appears in Farrington translated as follows:

THE MASCULINE BIRTH OF TIME OR, THE GREAT INSTAURATION OF THE *DOMINION* OF MAN OVER THE UNIVERSE.[55]

When one turns to the Latin in the Spedding, Ellis and Heath edition, however, Bacon's words in fact read:

TEMPORIS PARTUS MASCULUS SIVE INSTAURATIO MAGNA *IMPERII* HUMANI IN UNIVERSUM[56]

Where Farrington uses the English word 'dominion', Bacon in fact wrote '*imperii*'. Farrington's translation of *imperium* as *dominion* occurs not only in the title of Bacon's work but consistently throughout the text. In the first chapter, in a dialogue between a wise man and his young friend or son, Farrington makes the following translation:

> So may it go with me, my son; so may I succeed in my only earthly wish, namely to stretch the deplorably narrow limits of man's *dominion* over the universe to their promised bounds.[57]

Bacon's Latin text actually reads as follows:

> Ita sim (fili) itaque humani in universum *imperii* angustias nunquam satis deploratas ad datos fines proferam.[58]

Again in Chapter 2, Farrington's translation reads:

> But verily that would be to sin on the grand scale against the golden future of the
> human race, to sacrifice its promise of dominion by turning aside to attack transitory
> shadows.[59]

Bacon's exact words were:

> Nae ilud peccatum fuerit largiter in humani generic *fortunam auream*, pignus imperii,
> si ego ad umbrarum fugacissimarum insecutionem deflecterem.[60]

Here, had Farrington translated Bacon's words as 'man's *empire* over the universe', scholars relying upon his text might have considered the possibility
that Bacon was working with an idea of *empire* in this text, and that the idea
of man's original dominion, expressed here as 'empire', was part of a coherent
theme in Bacon's work.

The Spedding, Ellis and Heath edition of Bacon's works also avoided using
the term 'empire' when that was precisely the term Bacon used. Instead, in *De
Sapienta Veterum*, 'imperium' is translated as 'kingdom'. Again, my point is not
that this is an inaccurate translation of Bacon's meaning, but that had the trans-
lators used the term 'empire' in English, scholars would be more aware of the
significance of the concept of empire to Bacon's work. *De Sapienta Veterum* (*The
Wisdom of the Ancients*) deals with ancient myths, for which Bacon provides
an interpretation. We know that Bacon valued these myths, because he emphasizes in his Preface that he finds 'so close and so relevant connection between the
thing signified and the allegory'.[61] Moreover, Bacon intended the stories to be
instructive for his contemporaries: 'Though the subjects be old, yet the matter is
new; while leaving behind us the open and level parts we bend our way towards
the nobler heights that rise beyond.'[62]

Let us consider the Riddle of the Sphinx, which Bacon identifies as a riddle
about the nature of science. Put briefly the story is that in Thebes there lived
a sphinx who foisted riddles upon passers-by and would tear to pieces those
who could not solve them. The Thebans offered sovereignty over their city to
anybody who could solve one of the sphinx's riddles. Oedipus embraced the
challenge, solved a riddle and became ruler of Thebes. Bacon argues that the
fable is about science's 'application to practical life'.[63] Note the Spedding, Ellis
and Heath translation of Bacon's reading of the myth:

> [The riddles have] two conditions attached to them; distraction and laceration of
> mind, if you fail to solve them; if you succeed, a *kingdom*. For he who understands his
> subject is master of his end; and every workman is king over his works.[64]

Bacon's exact words are:

> Proinde in aenigmatibus Sphingis duplex semper proponitur conditio; non solventi mentis laceratio; solventi *imperium*. Qui enim rem callet, is fine suo potitur, atque omnis artifex operi suo imperat.⁶⁵

A few lines later:

> Now of the Sphinx's riddles there are in all two kinds; one concerning the nature of things, another concerning the nature of man; and in like manner there are two kinds of *kingdom* offered as the reward for solving them; one over nature, and the other over man. For the *command* over things natural, – over bodies, medicines, mechanical powers, and infinite other of the kind – is the one proper and ultimate end of true natural philosophy.⁶⁶

Bacon's Latin reads:

> Aenigmatum autem Sphingis duo in universum sunt genera; aenigmata de natura rerum, atque aenigmata de natura hominis: atque similiter in praemium solutionis sequuntur duo *imperia*; *imperium* in natural et *imperium* in homines: verae enim philosophiae naturalis finis proprius et ultimus est, *imperium* in res naturales, corpora, medicinas, mechanica, alia infinita; licet Schola, oblates contenta et sermonibus tumefacta, res et opera negligat et fere projiciat.⁶⁷

Later in the text, instead of kingdom, Spedding, Ellis and Heath use 'sovereignty'

> whence it follows that the Sphinx has the better of them, and instead of obtaining the *sovereignty* by works and effects, they only worry and distract their minds with disputations.⁶⁸

Yet Bacon used the same term as he did throughout the riddle, 'empire':

> unde fit ut (praevalente Sphinge) potius per disputations ingenia et animos lacerent, quam per opera et effectus imperent.⁶⁹

The first way that Bacon used the term 'empire' then, was to describe man's epistemic dominion. Bacon also used the word in a strictly political sense, to denote the power of the English monarch over his kingdom. As I noted earlier, this use of 'empire' stems from the Roman law tradition in which a king's power in his kingdom was described by the phrase *Rex in regno suo est imperator*,⁷⁰ and was used by Henry VIII in the 1533 Act in Restraint of Appeals. There was nothing remarkable about this sense of Bacon's use of 'empire'; it is simply important to note that, for Bacon, the Roman and Biblical traditions of empire did not seem to be in conflict.

Bacon's Roman use of empire occurs, for example, in 'Certain Articles, or Considerations, touching the union, of the Kingdomes of England, and Scotland'. Published posthumously, Bacon conceived of the union of the kingdoms in imperial terms: 'For the *Ceremoniall Crowns*, the Question will be, whether there shall be framed, one, new, Imperiall *Crown*, of *Britain*, to be used for the times to come?'[71] The political use of empire also occurred in many of Bacon's political writings, for example in the eighth book of *De Augmentis Scientiarum*, where the first chapter is titled 'The Division of Civil Knowledge into the Doctrine Concerning Conversation, Negotiation, and Empire or State Government',[72] in his essay 'Of Empire'[73] and in various letters to Elizabeth, James I and the Duke of Buckingham. This political and Roman tradition of empire has been explored in the literature on early modern sovereignty, and more recently by Armitage. What concerns us here is Bacon's use of the idea of Adam's empire of knowledge and the nature of its possible connection to the New World.

We know from Bacon's writing that there was a connection between epistemic empire and the New World because he frequently explicated the idea in the context of geographic expansion. In the *New Organon*, Bacon posits a cause–effect relationship between the opening up of the geographic world, and the opening up of new knowledge:

> By prolonged voyages and journeys (which have become prevalent in our times) many things in nature have been disclosed and found out which could shed new light on philosophy. And surely it would be a disgrace to mankind if, while the expanses of the material globe, i.e. of lands, seas, and stars, have in our times been opened up and illuminated, the limits of the intellectual globe were not pushed beyond the narrow confines of the ancients' discoveries.[74]

Several aphorisms later, Bacon repeats the analogy between geographic and epistemic expansion this time referring to the book of Daniel:

> We must not forget the prophecy of *Daniel* concerning the last ages of the world: that *Many shall go to and fro and knowledge shall be increased*, which manifestly hints and signifies that it was fated (i.e. Providence so arranged it), that thorough exploration of the world (which so many long voyages have apparently achieved or are presently achieving) and the growth of the sciences would meet in the same age.[75]

The aphorism illustrates Bacon's making reference to man's dominion of knowledge through a contrast between the New World and the ancient world. Comparing the present world with that of the ancients, Bacon writes:

> they knew but a tiny part of the regions and territories of the world, seeing that they lumped all the northern peoples together as Scythians, and all the peoples of the west as Celts; and knew nothing of Africa beyond the nearest part of Ethiopia, nothing of Asia beyond the Ganges, and still less of the provinces of the New World ... But in our times many parts of the New World and the furthest limits of the Old have become familiar in all respects, and the accumulation of experiments has grown beyond recognition.[76]

Bacon makes it clear that man's empire of knowledge is not something gained by philosophers ensconced in disputations far away from the affairs of the state. On the contrary, as Bacon explains in a manuscript note entitled 'In Praise of Knowledge', natural philosophy is a practical discipline:

> Printing, a gross invention; artillery, a thing that lay not far out of the way; the needle, a thing partly known before; what a change have these three made in the world in these times ... And those, I say, were but stumbled upon and lighted upon by chance. Therefore, no doubt the sovereignty of man lieth hid in knowledge; wherein many things are reserved, which kings in their treasure cannot buy ... their seamen and discoverers cannot sail where they grow. Now we govern nature in opinions, but we are thrall unto her in necessity; but if we would be led by her in invention, we should command her in action.[77]

The result of the close relationship between the growth of knowledge and overseas exploration is made explicit in the penultimate aphorism of the *New Organon*:

> Again consider (if you will) the difference between the life of men in any of the most civilized province of Europe and in one of the most savage and barbarous regions of the New Indies, and then you will think it great enough to justify the remark that *Man is a God to man*, not just for the relative benefits and helps they enjoy but also for their sheer material lot. And this difference does not spring from soil, climate or bodily constitution but from the arts.[78]

The reference to the 'New Indies' is not unique. In the *New Organon*, Bacon frequently refers to the idea of epistemic empire in the context of the discovery of the New World. In Book II, for example, he reports how '*Acosta*, and some others (after careful investigation), have observed that high tides happen on the coast of Florida at the same time as on the opposite coasts of Spain and Africa, and that low tides are also simultaneous'.[79] The connection between Bacon's ideal of epistemic empire and the New World compelled him to confront the question of how to utilize these newly discovered stocks of natural knowledge.

Bacon and Colonization

> For indeed, *Vnions*, and *Plantations*, are the very *Nativities*, or *Birth-Dayes* of *Kingdomes*. Wherein, likewise, *your Majesty* hath yet a Fortune extraordinary, and Differing, from former Examples, in the same Kind. For most Part of *Vnions*, and *Plantations*, of *Kingdomes*, have been founded, in the [e]*ffusion* of *Bloud*; But your *Majesty* shall build, in *Solo puro, & in Area pura*, that shall need no *Sacrifices Expiatory* ... or *Bloud*; And therefore (no doubt,) under a Higher, and more Assured, *Blessing*.[80]

In this passage from 'Certain Considerations Touching the Plantation in Ireland', Bacon tried with all his rhetorical might to persuade James I that plantations must be conducted honourably. The 'effusion of blood' was simply unacceptable. Bacon implored the king to 'build, in *Solo puro, & in Area pura*' (in a pure soil and in a pure space). Bacon's ideal of restoring mankind's empire over nature confronted him with a problem. If this empire of knowledge relied upon information gleaned in the New World, how would this knowledge be organized, administered and collected? Would colonization serve this purpose? Or would colonization instead cause bloodshed, greed and corruption? In this section I aim to provide a close and contextualized reading of Bacon's views on colonization. Far from being an advocate of colonization, Bacon held anxieties about colonization which meant that he did not consider colonies to be the right means through which man's epistemic empire should be restored.

Bacon approached colonization from a humanist perspective. The classical Roman anxieties about imperialism, found in Cicero's *De Officiis* and *Brutus*, as well as Sallust's *War with Cataline*, articulated moral concerns about the health of the Roman citizenry as the Republic and Empire respectively expanded into foreign territories. Writing as the Roman Republic collapsed shortly after the assassination of Julius Caesar, Cicero expressed a fear of the corruption created by foreign expansion and imperial power. In *De Officiis*, for example, he argued that 'there are many, especially those greedy for renown and glory that ... think they will appear beneficent towards their friends if they enrich them by any means whatsoever.'[81] Similarly, in Cicero's *Brutus*, a dialogue on the art of oratory, Cicero presents a genealogy of famous Roman orators, whilst instructing Atticus his interlocutor of the dangerous influence of Asiatic effeminacy on rhetorical style.

Like Cicero, Sallust's *War with Catiline* depicts the final years of the Roman Republic as a period of greed generated by a corruptive Asiatic influence: 'After the domination of Lucius Sulla, Catiline had been seized with a mighty desire of getting control of the government.'[82] This stemmed from the fact that 'in order to secure the loyalty of the army which he had led into Asia, he allowed it a luxury and license foreign to the manners of our forefathers ... he demoralized the warlike spirit of his soldiers.'[83] Bacon held similar concerns in relation to the English.

In a letter to the Duke of Buckingham, Bacon discusses 'Foreign Plantations and Collonies abroad'. He tells Buckingham that it is 'both honourable and profitable to disburthen the Land of such Inhabitants as may well be spared and to imply their labours in the Conquest of some Forreign parts without injury to the natives'.[84] Bacon then sets out some 'cautions to be observed in these undertakings'. He instructs Buckingham 'to make no extirpation of the Natives under pretence of planting Religion, God surely will no way be pleased with such sacrifices.'[85] Bacon's reference to preventing the dispossession and destruction of indigenous people is perhaps the most revealing of Bacon's anxieties about colonization. This was also the sentiment in Bacon's essay on plantations, that colonies must not dispossess indigenous people. They must be established in 'a

Pure Soile; that is, where People are not *Displanted*, to the end, to *Plant* in Others. For else, it is rather an Extirpation, then a *Plantation*.[86]

In an earlier work, presented to Queen Elizabeth, Bacon outlined his beliefs clearly: 'The Queen seeketh not an extirpation of that people, but a reduction; and that, now she hath chastised them by her royal power and arms, according to the necessity of the occasion, her majesty taketh no pleasure in the effusion of blood, or displanting of ancient generations'.[87] Bacon was consistent in his articulation of this anxiety. He warned against dispossession and annihilation not only in his advice to the English monarchs, but also to Robert Cecil and the Duke of Buckingham. To Cecil, he wrote, 'for if the wound be not opened again, and come not to a recrudency by new foreign succours, I think that no physician will go on much with letting blood'.[88] Similarly, in his advice to Villiers, Bacon argued 'make no extirpation of the natives under the pretence of planting religion: God surely will in no way be pleased with such sacrifices'.[89]

Instead of using violence to suppress the Irish, Bacon devised a plan for their naturalization. This involved the reproduction of English government and civic knowledge: 'If you *Plant*, where savages are, doe not onely entertaine them with Trifles, and Gingles, But use them justly, and gratiously, with sufficient Guard nevertheless ... send of them, over to the country that *Plants*, that they may see a better Condition than their own, and commend it when they return.'[90] This was not simply a moral policy. It was a fundamentally novel development of classical anxieties because Bacon was concerned with a policy of education and naturalization; with the recovery of the hearts – and minds – of the Irish. Elsewhere, he advocated a 'liberal proclamation of grace and pardon ... and ... a toleration of religion'.[91]

Bacon deliberately encouraged the English to rule their colonies justly in order to raise themselves above the barbarism of the Spanish conquistadors. The violent Spanish *encomiendas* and dispossession of the Amerindians met with dissent even among their own scholars, including Francisco de Vitoria and Bartolome de Las Casas. It is entirely possible that knowledge of these men's writings reached England. Regardless of whether Bacon had read Vitoria and Las Casas, he was certainly aware of the violence of the Spanish, which was at odds with his own classical ideal of government through the laws, and over men who were taught to reason and use knowledge. The intended contrast between the English and the Spanish is clear: the Spanish colonial policy of *encomiendas* relied upon the idea that the American Indians were barbarous slaves rather than reasonable men. I would suggest, therefore, that the best context for understanding Bacon's views on colonization is the context of Spanish violence in the New World. Where the Spanish used violence and dispossession, the English were to adopt a policy of granting civic laws to their colonies, and incorporating the indigenous people into the English Commonwealth. Throughout his work, Bacon advocated bringing just laws to the Irish. 'In countries of new populations', he wrote, one must 'invite and provoke the inhabitants by ample liberties and charters'.[92] He advocated incorporating the Irish into the Commonwealth, through 'the carrying of an even course between the English and Irish, as if they were one nation ... is one of the best medicines of that state'.[93]

The very fact that Bacon put forward a policy of naturalization reveals that he believed the Irish possessed qualities enabling them to emulate the English. In 1606, Bacon made a 'Speech in Favour of Naturalisation' to the House of Commons. Far from being brutes that needed suppressing through a system of *encomiendas*, the inhabitants of both Ireland and Scotland were 'ingenious, in labour industrious, in courage, valiant, in body, hard, active and comely'.[94] This is reminiscent of Tacitus's evocation of the virtues of the Germans in his *Germania*, a text in which he expressed anxiety about the Roman Empire through favourably comparing the virtues of the Germans with the questionable morality of the Romans.[95]

Naturalization, however, was a daunting task. Bacon was unsure whether it could be accomplished by the English when they simultaneously faced the difficulty of governing the frequently rebellious Irish, who regardless of their ingenuity, were also 'noted to be a people not so tractable in government'.[96] As we have seen, Bacon knew of the difficulties calming Ireland in the past:

> The Irish was such an enemy as the ancient Gauls, or Britons or Germans were, and that we saw how the Romans, who had such discipline govern their soldiers and such donatives to encourage them ... yet when they came to deal with enemies which placed their felicity only in liberty and the sharpness of the sword, and had the natural and elemental advantages of the woods and bogs and hardness of bodies, they ever found they had their hands full of them.[97]

Given the difficulties of governing colonies honourably, it was questionable whether maximizing territory was an advisable policy at all. In his advice to George Villiers, Bacon suggested that perhaps the territory over which the English king exercised empire was already large enough: 'And for a foreign war intended by an invasion to enlarge the bounds of our empire, which are large enough, and are naturally bounded with the ocean, I have no opinion either of the justness or fitness of it; and it were a very hard matter to attempt it with hope of success, seeing the subjects of this kingdom believe it is not legal for them to be enforced to go beyond the seas, without their own consent upon hope of unwarranted conquest.'[98]

What is most interesting here is Bacon's reference to England's territory being naturally bounded by the ocean. Bacon's support for the idea that the English should limit their territory is found elsewhere in his work. In 'Of the True Greatness of the Kingdom of Britain', for example, he confirmed that 'there is commonly too much ascribed to largeness of territory'.[99] In that tract, he set forth certain conditions for the expansion of territory, and commented that 'territories be compacted and not dispersed'.[100] Moreover, he argued that 'if the parts of an estate be disjoined and remote, and so be interrupted with the provinces of another sovereignty; they cannot possibly have ready succours in case of invasion, nor ready suppression in case of rebellion'.[101] This quotation epitomizes Bacon's complex, and at times internally discordant, views on colonization. In principle he believed in the virtue of establishing colonies, which explains his membership of the Virginia and Newfoundland companies. Yet the day-to-day practice of maintaining and governing colonies enmeshed the English in a number of moral problems. It is striking that Bacon makes no connection between colonization and the renewal of man's empire of knowledge. In fact, as we shall now see, when Bacon came to describe the ideal society that pursued the restoration of natural knowledge, he envisioned that society employing an explicitly non-colonial means of gathering information from the New World.

Bacon's Ideal of Empire

In Bacon's natural philosophy we find a non-colonial method of gleaning information abroad. This method involves the travels of experts and explorers who are itinerant and never settle or colonize the lands they encounter. It is not only in the *New Atlantis* that we find evidence for this method of collecting information; the rest of Bacon's work is dotted with references to information about places or natural phenomena abroad which have been collected by explorers rather than colonists. This disconnect in Bacon's work between epistemic and colonial empire is in marked contrast with the Hartlib Circle, Robert Boyle, the Royal Society and John Locke, all of whom saw colonization as an important means of supplying natural knowledge.

In Bacon's *Sylva Sylvarum*, an eclectic series of experiments of natural history, there are frequent references to information that has obviously derived from travellers' accounts of the Americas and the New World. 'In Peru, and divers parts of the West Indies', Bacon writes, 'the heats are not so intolerable as they be in Barbary, and the skirts of the torrid zone.'[102] This quotation derives from experiment 398 in the fourth Century of *Sylva Sylvarum*. The whole experiment is based upon information about Peru. Bacon has actually taken this paragraph almost directly from the Spanish Jesuit missionary Jose D'Acosta's *History of the Indies*, which was translated into English in 1604. In a later section of the *Sylva Sylvarum*, under the heading 'experiments in consort touching foreign plants', Bacon writes that 'it is reported that earth that was brought out of the Indies and other remote countries for ballast for ships, cast upon some grounds in Italy, did put forth foreign herbs, to us in Europe not known'.[103]

Jose d'Acosta, who travelled to Latin America in 1572 and remained until 1587, is often cited by Bacon, particularly in the *Historia Ventorum*. Under the heading 'General Winds', Bacon reports that 'persons sailing in the open sea between the tropics are aware of a steady and continual wind ... this wind is so strong ... it prevents vessels sailing to Peru from returning by the same way'.[104] These kinds of references occur every few pages. Bacon tells us, for example, that 'Peru ... may vie with Europe in the mild and temperate nature of the Air'. This is taken from Acosta's *Natural and Moral History of the Indies*, ii, 9.[105] Acosta's second volume also provides the information that 'with respect to the towns of Plata and Potosi in Peru' there is a great temperature variation between them, despite their similar geographic situation.[106] Under the heading 'Things Contributing to Winds', Bacon reports that 'Acosta observes that in Peru ... there is most wind at the full moon'.[107] Within the following two pages of that citation, Bacon cites Acosta's observations on Peru twice more.[108] He also refers to information about Peru and Chile on 'the limits of winds'[109] and other information that was apparently first discovered by Columbus.[110]

The references to American natural history are also extensive in Bacon's 'Historia Vitae et Mortis' ('The History of Life and Death'). Here Bacon discusses medicines that make opiates, and these include the 'Indian leaf'.[111] He also reports the longevity of the native people of 'Brazil and Virginia',[112] and a drink made with 'Indian corn'.[113] In the translation of the 'De Fluxu et Refluxu Maris', Bacon's writes at length about the 'motion and flow of waters from the *Indian* Ocean to the *Atlantic*, a motion which is swifter and stronger towards the Straits of *Magellan*'.[114]

All these references indicate that Bacon relied upon information sourced in the Atlantic. He knew well that there was an intimate connection between knowledge and the New World, but he did not conceive of any relationship between *colonies* and the collection of knowledge. This was an innovation that the Hartlib Circle would make in the 1640s, as the next chapter will show. Bacon's ideal method of gaining information in the Americas was to rely upon correspondents and travellers. He explains the method eloquently in his *Parasceve*, or *Preparations Towards a Natural and Experimental History*:

> The materials for the intellect are so widely spread out that they ought to be sought out and gathered (as if my gents and merchants) from all sides. I think too that it is rather beneath the dignity of my enterprise to spend my own time on a matter which is open to practically everyone's efforts.[115]

Bacon specified a number of 'injunctions' to travellers in his 'De Fluxu et Refluxu Maris', outlining the kind of knowledge he desired from the New World: 'Inquire whether the time of high tide on the shores of *Brazil* precedes the time of high tide on the shores of *New Spain* and *Florida*'.[116] I would suggest that Bacon's discussion of the means of collecting natural knowledge from the New World provides a useful context for understanding his utopia, *New Atlantis*. The College of the Six Days' Works is an institution based upon precisely this type of New World epistemic adventure.

The utopia is an incomplete, highly symbolic narrative of Spanish sailors who become shipwrecked off the island of Bensalem, home to a scientific community entitled the College of the Six Days' Works. The dominant reading of *New Atlantis* sees no anxiety about colonization. Rather, Bensalem is seen as legitimating colonial possession. Denise Albanese, for example, argues that the work is 'ideological'. This, she writes 'enables the scientific to be produced under the ostensible sign of the literary, and results in an act of discursive colonization that tropes the cultural agendas of Jacobean imperialism'.[117] In a similar vein, Amy Boesky sees the utopia as one of the 'founding fictions' which underlies colonization,[118] and Charles Whitney sees science as a metaphor for colonization in his well known article entitled 'Merchants of Light: Science as Colonization in the *New Atlantis*'.[119] Recently, however, two historians have argued that Bacon's utopia does not legitimate colonial possession. Claire Jowitt suggests that 'the scientocracy described in the *New Atlantis* is able to care for all the population's needs without territorial expansion or foreign trade'.[120] In an article on Bacon's natural knowledge in the *New Atlantis*, Richard Serjeantson comments that the text is 'pre-colonial' because it depicts the world as 'still an imperfectly known and unexplored place'.[121]

The *New Atlantis* is not so much pre-colonial but anti-colonial. A society sending out its citizens to establish a colony was a prominent *topos* in the tradition of early modern utopian literature and featured, for example, in Thomas More's *Utopia* (1516).[122] Bacon was, as we have seen, acutely conscious of the issues surrounding colonization. Given that he believed in the symbiotic opening up of the geographic and intellectual worlds, there is no reason why Bacon would have wanted to ensure that his utopia was set in a *pre*-colonial moment, as this would mean that Bensalem predated even ancient colonizing societies, notably the Romans. Given that the Bensalemites are Christians, it is highly unlikely that the society is simultaneously pre-colonial.

A more plausible explanation for Bacon's silence on colonization, I think, is that the text is anti-colonial. By imagining a society that does not engage in colonization, Bacon implies its redundancy. Moreover, the utopia presents an alternative, non-colonial, ideal of encounters between Bensalem and foreign lands:

> Every twelve years, there should be set forth out of this kingdom two ships, appointed to several voyages ... on a mission of three of the Fellows or Brethren of Salomon's House; whose errand was only to give us knowledge of the affairs and state of those countries to which they were designed, especially of the sciences, arts, manufactures, and inventions of all the world; and withal to bring us books, instruments, and patterns in every kind; That the ships, after they had landed the brethren, should return; and that the brethren should stay abroad till the new mission ... Now for me to tell you how the vulgar sort of mariners are contained from being discovered at land;

and how they that must be put on shore for any time, colour themselves under the names of other nations ... I may not do it ... But thus you see we maintain a trade, not for gold, silver, or jewels; nor for silks; nor for spices; nor any other commodity of matter; but only for God's first creature, which was *Light:* to have *light* (I say) of the growth of all parts of the world.[123]

Far from remaining to establish a colony, the ships 'should return', and interestingly, 'the brethren should not even let their presence be known by the Europeans'. The enterprise is described not as colonization, but as a 'trade'. Bensalem's self-sufficiency enabled Bacon to put forward an ideal of the collection of knowledge through explorative travels. These trips would be specially targeted to 'give us knowledge of the affairs and state of those countries to which they were designed, especially of the sciences, arts, manufactures, and inventions of all the world; and withal to bring us books, instruments, and patterns in every kind'.[124] What is utopian about Bensalem is that it fulfils the promise of Bacon's project for natural knowledge whilst avoiding the ethical dangers of colonization.

In a number of ways, the method of collecting information in Bensalem is consistent with Bacon's views on the proper pursuit of natural knowledge elsewhere in his work. First and most importantly, the College of the Six Days' Works is so called because it aims to restore the natural knowledge which man possessed just after the creation:

> this Order or Society is sometimes called Salomon's House and sometimes the college of the Six Days Works; whereby I am satisfied that our excellent king had learned from the Hebrews that God had created the world and all that therein is within six days; and therefore he instituting that House for the finding out of the true nature of all things.[125]

Furthermore, this aim is expressed in terms of the idea of man's epistemic empire. The Father tells the sailors that 'the End of our Foundation is the knowledge of Causes, and secret motions of things; and the enlarging of the bounds of Human Empire, to the effecting of all things possible'.[126]

The second way in which Bensalem's production of natural knowledge is consistent with Bacon's general principles is the emphasis upon practical knowledge.[127] Salomon's House is concerned, for example, with the prolongation of human life. 'We have dispensatories, or shops of medicines', 'we have perspective houses, where we make demonstrations'.[128] Moreover, this knowledge is produced through inductive experimentation. The Father of Salomon's House tells the sailors:

> We have three that bend themselves, looking into the experiments of their fellows, and cast about how to draw out of them things of use and practice for man's life, and knowledge as well for works as for plain demonstration of causes ... Lastly we have three that raise the former discoveries by experiments into greater observations, axioms and aphorisms. These we call Interpreters of Nature.[129]

Mary Poovey and others have pointed out that Bacon was concerned with issues surrounding the production and reliability of knowledge and that his emphasis upon inductive experiment should be seen in this light.[130] I would suggest that we should place the *New Atlantis* in this context, and then view the text in light of our understanding of Bacon's colonial anxieties. From this reading, Bensalem emerges as a society that deals with the issues of the production and reliability of knowledge gleaned in the New World, and intentionally avoids the issue of colonization. The *New Atlantis*, with its travellers gleaning information for the collective pursuit of natural philosophy, is Bacon's solution to the problem of knowledge.

Advocating the restoration of man's empire over nature, but sceptical about colonization as its means for achievement, Francis Bacon is an illuminating starting point for investigating the intellectual origins of the British Empire. Bacon's natural philosophy epitomizes the idea of epistemic empire, and shows that its origins in a Biblical tradition of Adam's empire were independent from any tradition of colonization. In the early seventeenth century, therefore, there was no necessary connection between man's prelapsarian empire and colonization. Yet half a century later, for Bacon's proud successors in the Royal Society of London, and for John Locke, the two were intimately connected. What ideational moves made this possible? This book tells the story of the coming together of man's empire over nature, and England's empire of colonies.

2 RESTORING EDEN IN AMERICA: THE HARTLIB CIRCLE'S PANSOPHICAL EMPIRE

To the Virginian Gentlemen Planters.

Sirs, what's to your Eye and Eares presented,
Is for your Honour and Wealth intended
If now a Virgins counsel you will take,
Great Treasure of it you shall surely make
And if more South a little you will go,
Infinite Riches shall upon you flow.
Adde to't a Westerly Discovery
Then happy are you made eternally.

Virginia Ferrar, *A Rare and New Discovery ... For the Feeding of Silk-worms* (1652)

The final pages of Virginia Ferrar's pamphlet heralding the 'rare and new discovery' for the 'feeding of silk worms' are a poetic tribute to the Virginia colony and its agrarian fruitfulness. What Ferrar admired about the silkworm was its utility. Properly cultivated, the silkworm would yield 'great treasure' which would be to the advantage of both England and her colonies.

Virginia Ferrar (*c.* 1627–88), presumably named after the Virgin Queen and the eponymous American colony, was the daughter of John Ferrar (*c.* 1588–1657), a deputy to the Virginia Company and a member of four other colonial companies.[1] He owned a large estate in Virginia and published eight pamphlets encouraging planters in the Virginia colony to cultivate silkworms. Ferrar also owned a manor, Little Gidding in Huntingdonshire, where his daughter Virginia kept silkworms and wrote poetry extolling the virtue and utility of silk cultivation. The Puritan intelligencer Samuel Hartlib (*c.* 1600–62) supervised the publication of 'The Reformed Virginian Silkworm' in 1655.

Most scholarship on Samuel Hartlib and his circle centres upon their Puritan religious convictions, their sprawling correspondence and their various pedagogical and social reform programmes.[2] Virginia Ferrar's poem about silkworms, however, draws our attention to a neglected aspect of the Hartlib Circle's endeavours. This is their extensive work on planting, and the relationship of their natural philosophy to English colonial ventures in the Atlantic.

It is my contention that colonization was a principal intellectual foundation of the Hartlib Circle's natural philosophy, and that they developed their ideas about agrarian cultivation in the context of English colonization in the Atlantic. Moreover, colonies provided the ideal place for the improvement of land: they were experimental spaces where the world could be reformed. While the scholarship recognizes that the Hartib Circle were interested in colonization, it is usually assumed that colonies were a blank slate to which natural philosophy could be applied. As Mark Greengrass, Michael Leslie and Timothy Raylor put it, 'Universal Reformation ... could camouflage the colonizing aspirations of the English in Ireland.'[3] The assumption is that the process was one way. I hope to show, however, that the process was reciprocal. Colonization played a formative role in the Hartlib Circle's natural philosophy, an influence we see most profoundly in William Petty's work on Ireland. The exigencies of colonizing produced a type of natural philosophy which was increasingly experimental and quantitative.

It is by no means unproblematic to speak of the Hartlib Circle as an entity. Hartlib's correspondents came from different countries and socio-economic classes, and they spoke different languages. The Circle was not an official organization; it was a loose affiliation of men with diverse interests. Keeping in mind this caution against reification, I do think it is appropriate and useful to treat Hartlib's circle of friends as an entity because of their interest in using husbandry as part of their endeavours to reform the world.[4]

Despite their diverse backgrounds, as J. Peacey has pointed out, among the Hartlib Papers there is a document which suggests that a number of Hartlib's associates in a broader group of men were united by colonial business interests. The twenty-five men were listed as 'preparing to contribute to the costs of writing and translating, as well as printing and distributing, Puritan works.'[5] At least thirteen of the men were members of the Massachusetts Bay Company, and a number were also involved in the Providence Island Company.[6]

A second note of caution must be sounded on the issue of using the term 'natural philosophy' to characterize the group's work. It could be argued that it is more accurate to describe the group's work as agricultural improvement or husbandry, but for a number of reasons I think it is reasonable to use 'natural philosophy', albeit cautiously, as an umbrella term to describe the kind of natural knowledge the group pursued. First, Hartlib and his close associates saw themselves as building upon a number of Bacon's aspirations for the practical application of knowledge, and they also shared Bacon's belief in the project of using natural knowledge to restore man's original dominion over nature. Moreover, Hartlib was a mentor to Henry Oldenburg, who became Secretary of the

Royal Society. These intellectual and personal associations place the Hartlib Circle in the Baconian tradition of seventeenth-century natural philosophy.[7]

The second reason for using the term natural philosophy is that this is the term Hartlib himself frequently used to describe his intellectual enterprise. In his work *The Compleat Husband-Man*, for example, Hartlib introduces letters from the Americas that he copies out, describing them as '*Copies and Extracts of more letters written to Mr.* Samuel Hartlib: *They all tending very much to the great improvement not onely of* Agriculture, *but of true and real* Learning, *and* Naturall Philosophy'.[8]

Importantly, the Hartlib Circle developed a new dimension of the Baconian project. They emphasized the agrarian aspect of the Fall, and thus the agrarian nature of any redemptive labour to restore man's original empire over nature. It was husbandry – the cultivation of the earth for the benefit of mankind – which played the leading role in the Hartlib Circle's programmes for reformation. The significance of their interest in planting was not simply that it was part of a greater concern with the idea of 'improvement'; that is, methods of generating wealth through improving England's agriculture. This it was. But planting was also intimately connected to colonization. In fact, in the contemporary idiom, the two words were synonymous; plantations were often referred to as colonies, and vice versa. These linguistic connections should not surprise us, as colonization was conceived of as an agrarian enterprise. Ideas about husbandry, then, were well suited to be developed in the colonial context.

In the mid seventeenth century, the English intensified their colonial and commercial expansion in the Atlantic and Caribbean. The decades of the 1630s to 1650s witnessed the settlement of a number of colonies. Barbuda, Antigua and Montserrat, for example, were colonized in 1632, and Montserrat subsequently became the destination for Irish Catholics who flocked to the island after having served as indentured servants to the English in the other Caribbean colonies.[9] Cromwell's victory in Drogheda in 1649 precipitated a further influx of Irish, but this time they were prisoners. William Penn, Cromwell's commander in the Caribbean, wrested Jamaica from the Spanish in 1655. Back in mainland North America, Puritans from Massachusetts established themselves in Connecticut; formal government was installed on the island of Acquidneck (later Rhode Island) in 1639; and Massachusetts Bay absorbed the colony of Maine in 1652.

Meanwhile, England underwent something of a commercial revolution. The importance of foreign merchants declined as England increasingly generated sufficient numbers of merchants and trading posts to render London a centre of world trade.[10] With the introduction of the Navigation Acts from 1651, England was able to greatly restrict foreign shipping and ensure that the raw materials from England's colonies were traded for her own benefit, thus pressuring the Dutch who had dominated the carrying trade.[11]

During the 1630s, Samuel Hartlib proposed Virginia as an alternative place for the establishment of the godly utopia Antilia, an ideal community discussed by members of a Protestant fraternity in Elbing (Elblag), a Baltic town now part of Poland, which was Hartlib's birthplace around the turn of the seventeenth century. Hartlib returned to Elbing briefly after studying at the University of Koenigsberg, and then in Cambridge as a student of John Preston, the master of Emmanuel College, in 1625–6. Back in Elbing, Hartlib met the Calvinist minister John Dury, who became one of his closest friends.

Hartlib was compelled to flee Elbing for England in 1628 after the capture of Mecklenberg and Pomerania by the Habsburg armies in the Thirty Years' War. Once in England, Hartlib's first project was the establishment of an academy at Chichester in Sussex in 1630. It promptly failed and Hartlib retired to London where, during the next part of that decade, he established a manuscript service providing intelligence to the exiled English Protestant elite dispersed through continental Europe. Calling himself an 'intelligencer', Hartlib set himself up as a human hub of information. John Dury became one of Hartlib's first correspondents, as did the Dutch brothers Gerard and Arnold Boate who would later write a natural history of Ireland. Gerard was a physician and naturalist closely associated with the Boyle family. Arnold was also a physician and worked in Ireland under the surveyor-general, Sir William Parsons.

It was during this period in London that Hartlib began to correspond with men who were either born in or travelled to the Americas. These included the Bermudan-born alchemist, George Starkey; the natural philosopher and Governor of Connecticut, John Winthrop Jr; the Virginian Governor Edward Digges;[12] the merchant and member of the Irish, East India, Somers Islands and New Rivers Companies as well as deputy to the Virginia Company, John Ferrar; his daughter Virginia Ferrar whose work on silkworms Hartlib published; and the physician and traveller to New England John Child, who later came to work with Gerard Boate on the project of surveying Ireland.

In the 1630s, Hartlib began corresponding with Jan Amos Comenius (Jan Amos Komensky), who was a Czech theologian and a member of Unitas Fratrum (the Community of Brethren). Like Hartlib, Comenius was displaced by the Thirty Years' War. After completing his studies at the University of Heidelberg, Comenius moved with some expatriates to Leszno in Poland where he became a teacher at the local *Gymnasium*, a Germanic grammar school. Comenius's major project was the creation of a universal education system and, like Hartlib, he was influenced by the millenial hope that through reforming his knowledge of the world man might unite his knowledge of nature with his worship of God. Comenius termed this idea *Pansophia*[13] and articulated it best in his *Reformation of Schooles* (1642) and *The Great Didactic* (1657).

Hartlib became Comenius's dedicated publicist and set about persuading Comenius to join him in London. Upon his arrival in 1641, Comenius stayed with Hartlib for eight months. He was, however, disappointed with England's inadequacy as a place of rest from the turbulence of his homeland. In 1642, three months before he left England, Comenius, Hartlib and Dury signed a pact promising to devote themselves to 'religious pacification, education and the reformation of learning'.[14]

During the Civil Wars, Hartlib and Dury remained in London and unofficially aided the parliamentarians. After their victory in 1646, Hartlib attempted to establish an 'Office of Address' in London which aimed to promote the exchange of information. The hope was that the dissemination of knowledge would hasten the arrival of the millennium. As the book of Daniel put it, 'knowledge shall be increased' in the final days of the earth (Daniel 12:4)

Two years later a young physician named William Petty published an important manifesto addressed to Hartlib on the topic of the advancement of learning. Educated at Oxford, Amsterdam, Leiden, Utrecht and Paris, Petty was back in London by 1646 and well on the way to establishing a successful private practice. He was made a Fellow of the College of Physicians and was elected as a reader at Gresham College. Petty's status continued to rise with the victories of the parliamentarians. In 1652 he was appointed the physician to the army in Ireland. It was in this position that he became involved in the survey of the survey of captured Irish land, criticizing the method employed by the surveyor-general of Ireland, Benjamin Worsley, and putting forward his own survey in its place. Petty's method was rooted in calculative assessment of the colony through measurement and observation of the topography, population and trading potential of each area. Petty did well from his Irish exploits, managing to take possession of over 19,000 acres of Irish land.

As Charles Webster points out, the Hartlib Circle's agrarian projects were part of a Puritan attempt to reform the world, influenced by the millenarian sentiment of the book of Daniel which 'relate[d] to the opening up and exploitation of the New World'.[15] Certainly, Puritanism is one factor in the Hartlib Circle's interest in the New World, but it does not explain the significance of the Atlantic world for the Hartlib Circle's natural philosophy. If it did, we would expect the majority of the Hartlib Circle's correspondence and papers to be related to Puritan New England. This is not the case. From the Edenic fruitfulness of the Caribbean colonies, to the sugar industry in Barbados, the agricultural techniques of Virginia, or the plantation of Ireland, the rest of the Atlantic and Caribbean world figures as much in the Hartlib Circle's papers as does Puritan New England. The Hartlib Papers' sheer abundance of references to the Atlantic world as a whole is one fact which demands we broaden our interpretive context. Placing the group of natural philosophers in the context of English colonization in the Atlantic enables us to see that the Hartlib Circle did something conceptually significant. They brought together the idea of Adam's dominion over nature with colonization.

Restoring Eden in America

> I am apt to believe, that when God set Adam in the Garden Eden to keep it and dresse it, He meant [man] to exercise his Industry, as well about the discovery of the fruitfulnesse of perfect nature, which could not be without much delight to his understanding, as about the pleasantnesse of the place, which he could have by dressing increased, and made completely answerable to the perfection of his own imagination ... although we now come farre short of that knowledge, which he had in Nature ... by reason of that Curse is shut up unto us: Yet we find by Experience ... that in your Apiary in the Country, you not only found profit ... but that besides the benefit of Wax and Honey, you gained more delightful Observations of their Working and Government.

> Samuel Hartlib *The Commonwealth of Bees* (1655)[16]

Planting and husbandry were the primeval pursuits of man, the virtuous tasks of Adam in Eden. The Hartlib Circle's natural philosophy was inspired by an ideal of restoring the earth to its Edenic state of paradise. One of the books Hartlib owned explained the necessity of this Edenic pursuit particularly eloquently. '*God placed* [Adam] *in* Eden *not only to enjoy but to* labour *without both which no place can be a paradise ... Nor can there be found in* Nature *a more ingenuous, necessary, delightful, or honourable* employment *than agriculture; a* Calling *born with us.*'[17]

These words were written by Adolphus Speed, the English agriculturalist who corresponded with Hartlib, and whose debts Hartlib paid in 1650. Under the auspices of Sir Cheney Culpeper, but at Hartlib's request, Speed was retained in the Leeds Castle estate to advise Hartlib on rabbit-farming.[18] Speed's tract entitled *Adam out of Eden* (1659) argued for the revival of the Adamic pursuit of husbandry and suggested fifty-four schemes to aid the process. 'The advancement of all kinds of knowledge', Speed wrote, 'may rayse our expectation of more than a restauration to that naturall perfection, which hath beene since the fall of Adam.'[19]

Like many of the authors whom he read, Hartlib saw husbandry as the foundation of natural philosophy, a claim he made explicitly in *An Essay for the Advancement of Husbandry-Learning*. Husbandry, he claimed, was not only a 'Science', but also 'the Mother *of all other* Trades *and* Scientificall Industries'.[20] A husbandman was a farmer; he managed his domestic economy and cultivated his land. Husbandry was also what went on in colonial plantations because it was the process of improving land, rendering it fruitful and producing wealth for the good of England. As Patricia Seed has shown in *Ceremonies of Possession*, both foreign and domestic land could be brought under control in this manner, which was analogous to that of establishing a garden by enclosing and cultivating land.[21] The abundant potential for planting in the New World meant that the American plantations played the central role in the Hartlibians' vision. America was a potential Eden; a cornucopia of plants and animals and a land uncultivated by a people who had not yet learnt the word of God. The English colonies in America were seen as the right place to begin the recovery of Adam's empire.

The role that agriculture could play in sustaining and improving communities occupied a good deal of the Hartlib Circle's writings and correspondence. Hartlib made notes from books he read pertaining to the husbandry and improvement of American colonies. In 1649, for example, his notebook *Ephemerides* contains a description of 'Bullock's Booke of Virginia',[22] otherwise known as *Virginia Partially Examined* by William Bullock.[23] Its aim was to encourage Virginian colonists to improve their government and economy. William Bullock's father lived in Virginia for about a decade, and William took care of his estates and encouraged potential planters to settle in Virginia.[24] Hartlib's notes outline the book's argument that 'every body may bee made rich' and how 'the State may bee made rich or inrich it*self*'.[25] This could be achieved, Bullock argued, through innovation with respect to agricultural and mineral production. Indigo, iron, 'corne [which] is sown in March and reaped in Iune' were all potential resources.[26]

Another avid reader of Bullock's book was Benjamin Worsley, a physician who first came into contact with Hartlib in the 1640s when they were both members of the Invisible College, an association of like-minded utopian puritans.[27] Worsley developed a keen interest and involvement in colonies and their administration. He went on to become the secretary and treasurer of the Council of Trade and Foreign Plantations in 1672. Back in 1649 he had made a concise summary of all grants and charters that the Virginia colony had ever gained. He also devised a plan for the improvement of the English economy based upon colonial trade.[28]

In a letter to John Dury four years after the publication of William Bullock's book, Worsley described it as providing 'very good hynts' for policies to reform Virginia.[29] Unlike the Puritan New England colonies, Virginia was seen as morally depraved and led by a corrupt, disorganized local administration, reflecting the financial crisis in which the Virginia Company found itself in the 1620s.[30] When Worsley became secretary to the Council of Foreign Trade and Plantations he was continually involved in attempts to reform Virginia, as a number of letters to John Dury and the politician Walter Stickland attest.[31] As Peter Thompson has shown, Bullock's tract was one of a number of seventeenth-century works which 'attempted to justify their interest in what might be termed the poetics of husbandry by providing workable suggestions for improved agricultural practice grounded in empirical proof'.[32]

Husbandry was necessarily a collective enterprise and productive of fruit for the common good. In a letter to Hartlib, John Beale expressed his desire to note down his experiments so that they could be pursued by other like-minded men: 'Every day new Experiments grow upon me for which cause I think best to contracte the substance of itt into letters & to ingrosse every occurrence into larger volumes for the Common people. When they shall take a fire & be incouraged by leading Examples to dresse our English soyle into a goodly faire Paradise.'[33]

The idea of recreating paradise may seem utopian to the modern reader, but the Hartlib Circle's references to the ideal society were intended as realistic and practical. A perfect illustration is *A Description of the Famous Kingdome of Macaria*, which was written by Gabriel Plattes but published by, and till recently attributed to, Hartlib. Macaria, the ideal society, is organized to promote husbandry, plantation and colonization: 'They have five under Councels; to wit: A Councell of Husbandry; A Councell of Fishing; A Councell of Trade by Land, A Councell of Trade by Sea; A Councell for New Plantations.'[34] A traveller, describing the nature of the society to a scholar with whom he is in dialogue, tells him that 'The Councell of Husbandry hath ordered, that the twentieth part of every mans goods that dieth shall be employed about the imrpvoing of lands, and the Making of High-wayes faire, and bridges over Rivers; by which meanes the whole Kingdome is become like a fruitfull Garden'.[35] Moreover, men are sent out annually to establish plantations: 'In the Councell for new Plantations there is established a law, that every yeere a certaine number shall be sent out ... provided for at the publicke charge.'[36]

It was not only the Hartlib Circle's utopian writings which accorded husbandry a role in restoring the world to its Edenic state. Most of the group's other works on husbandry stipulate the role of their schemes in the improvement of the world. Like *Macaria*, these works have a powerful colonial dimension. In *A Loving Advertisement to all the Ingenious Gentlemen-Planters in Virginia now upon the Designe of Silk*, which Hartlib published in 1655, Virginia Ferrar hopes that the Virginian planters'

> noble courteous examples will allure all other Gentlemen Travellers to cast into this good work some mites of their further knowledges, and every man to contribute his prayers and help to this or any other hopefull designe: seeing the consequence of them may be so good and great, not only to the *English* Nation ... but to the poor *Savages* their welfare of souls and bodies, which God grant.[37]

'The Reformed Virginian Silkworm' contains a number of poems written about the Virginia colony. There is also a poem entitled 'From a *Virginia*-Planter in *England*, to the Virgin-Lady *Virginia*'. It lauds the rediscovery of hidden arts:

> *Amongst these Lands which sing the memory*
> *Of their deare children (who with pious care*
> *Have them ennobled by th'utility*
> *Of Arts, that long unto them hidden were)*

Virginia *boasts thy name in happy houre,*
Who to her Garland add'st so rich a flower

... ...

The glorious Silkwormes (of all Wormes the best) ·

...

'*Our lazy Clime, and those that sail from hence*

To blesse VIRGINIA ...

...

A Virgin *though,* VIRGINIA *is thy Name*
Which shall unto VIRGINIA *adde more Fame;*
For thou exalt'st her Honour by thy Hand
And mak'st the Silkwormes famous in that Land.[38]

The tract is addressed to Virginian planters, and states:

> this new invented way of thus keeping Silkwormes on the Trees; it requiring neither
> skill not pains ... that when the Indians shall behold and see you begin the businesse,
> they will with all alacrity set upon it likewise, and imitate you ... And thus by the
> blessing of Almighty God, there may be good hope of their civilizing and conversion,
> so that they may be likewise great gainers both in body and soul by this thing. And if
> this prove so ... how much then indeed will *Virginia's* happinesse be every way raised
> to the height of Blisse.[39]

In addition to cultivating the earth, therefore, the production of silk would be
beneficial to planters and Indians alike. As Ferrar wrote later in the text:

> it will be good for you to incourage the Savages, when they finde any bottoms in the
> woods, to bring them to you ... Now the two Propositions that tend to infinite wel-
> fare, benefit, and wonderfull advantage both to *England* and the Colony jointly are ...
> first ... that some kinde of Coyne be sent to *Virginia*, that may be authorized to passe
> there for their Commwece and better trading ... the second thing is, that a Publica-
> tion be procured and sent to the Colony in *Virginia*, declaring unto them, that there
> shall be livery for all men to bring from thence for the space of ten years to come, any
> commoditiy that they shall there raise, into *England*, Custome and Excize free.[40]

The tract concludes with the following hope:

> There remains nothing but humble thanks to God, and to these Gentlemen your due
> respects, whom God hath made such Instruments for your happiness, hoping their
> noble courteous examples will allure all other Gentlemen Travellers to cast into this
> good work some mites of their further knowledges, and every man to contribute his
> prayers and help to this or any other hopeful designe: seeing the consequence of them
> may be so good and great, not only for the *English* Nation at home and abroad.[41]

The reference to the 'English nation at home and abroad' indicates a second
significance of the Americas for the Hartlib Circle's project. Not only was the
New World the location for recovering Eden, but the American colonies would

also aid England's struggle against the Catholic powers by providing a source of wealth. In 1658 the Church of England clergyman John Beale wrote to Hartlib about the necessity of colonies for the advancement of the world: 'Wee should with mutuall endevours set our hearts to advance them; all true-hearted old Puritans to advance Newe England: all true Parliamentarians together with the men of the olde forme to advance Virginia & and her confiners; & our Protector to advance Iamaica, as the concernment of his greate fathers honour.'[42]

In the same vein, Benjamin Worsley's *Profits Humbly Presented* makes the argument that establishing overseas colonies will enable England to be self-sufficient, rather than having to rely upon trade with Spain, Italy and Turkey:

> There are severall commodities both of necesitie and pleasure *that* wee are forced to fetch out of other Countries ... by such a well regulated plantation ... the most generall of those commodities wee now fetch from other parts, may bee had within our own dominions and that at very inferiour rates especially those Countries we will plant being also ordered and improved, according to *the* former manner, & government of our praescribed Hussbandrie.[43]

Worsley was outlining how husbandry would aid English colonization. He explained that '<these> Commodities will bring into the Planter, Sayler, Merchaant and so to the whole Kingdome, most of all the Bullion in those partes ... because wee are able now, to afford them much cheaper'.[44] Then, Worsley points out, 'this enlargement of our Dominions & this encrease & improvement of our Nationall trading, will much spread *the* glory & add to *the* power & strength of this Kingdome'.[45]

Natural Philosophy and the Atlantic Colonies

There were two tangible ways in which the Atlantic world influenced the Hartlib Circle's work. First, the Atlantic colonies provided an important space for carrying out the reform of knowledge. Adam's original dominion over the earth was made possible by his perfect knowledge of the natural world. This omniscience was illustrated by his ability to give correct names to all the animals: 'Whatsoever Adam called every living creature, that was the name thereof' (Genesis 2:19). Hartlib and his colleagues believed in a mutually beneficial relationship between knowledge and husbandry: restoring knowledge of nature would enable man to develop the kind of agriculture and husbandry necessary to reap the world's fruitfulness. Conversely, husbandry and planting were productive of knowledge. Husbandry and natural knowledge coalesced in an enterprise that would benefit Protestant England. As Mark Greengrass, Michael Leslie and Timothy Raylor point out, in the work of the Hartlib Circle 'the perception of knowledge as a public commodity became of central concern for the first time'.[46]

Comenius's interpretation of this idea was what he called a pansophical reformation, which denoted a holistic reformation of knowledge. The essence of pansophical reformation was to produce an encyclopedia of useful knowledge, so to improve the state of humankind. Comenius published a number of textbooks for his reformation of knowledge. In the *Via Lucis*, written during his travels through Europe prior to 1641, he stipulated the invention of an artificial, universal and perfect language. Though the tract was not published until 1668 it circulated before then in manuscript form in England and Europe.[47] Comenius's most famous textbook was the *Janua Linguarum Reserata* (*The Gateway of Languages Open*) which instructed the teaching of Latin. It was this book that, as Robert Young has shown, the New England missionary John Eliot (1604–90) possessed; a Native-American's name was inscribed upon its front page.[48]

Comenius inspired various attempts to instigate education and language schemes in America.[49] John Eliot made a voyage to Boston in 1631. He attempted to proselytize to the Native Americans by translating the Bible into Algonquian, and from 1650 he established a number of 'praying towns' in which Native American Christians would live together.[50] Similarly, the missionary and founder of the town of Providence, Roger Williams, was influenced by Comenius. Williams arrived in Massachusetts in 1631, soon moved to Plymouth but was dismissed and moved again, this time to the area of Narragansett Bay after his views clashed with the church's established community. He began to build strong ties with native leaders and published *A Key into the Language of America* (1643) which was strongly influenced by Comenius's views.

In the writings of other members of the Hartlib Circle there was a clear connection between pansophical knowledge and planting; the study of husbandry and plantations would help create a dominion of knowledge because it revealed the secrets of nature. Experiments such as those suggested by Benjamin Worsley and copied out in Hartlib's *Ephemerides*, were 'of the highest Philosophy'. They were designed to produce the knowledge that would enable men to 'make out of Apples very good sugar which would bee to turne England into Barbados'.[51] This could not be achieved without knowledge of planting and husbandry. John Beale wrote to Hartlib in 1659 about the very practical agricultural matter of apple cider, and understood well the greater purpose of his experiment: 'Thus you may discourse your Cider into an Accomodable Philosophy, That will converte our whole land into Paradyse.'[52] Similarly, in his *Discoverie for Division or setting out of Land*, Hartlib explains that he wants to 'procure for the Publick Good a more full discovery of this Subject, which one who is a great searcher into the mysteries of Nature hath an inclination to write of'.[53] He wants this information 'in due time' to be 'imparted into this Common-wealth'.[54]

These sentiments are echoed throughout Beale's rather extensive collection of writings on agricultural topics, including *Herefordshire Orchards, a Pattern for All England* (1657), *Observations on some parts of Bacon's natural history, as it concerns fruit trees, fruits and flowers* (1658) and *Nurseries, Orchards, Profitable Gardens and Vineyards Encouraged, Advocating the Greater Availability of Timber and Fruit Trees* (1677).

When we turn to Hartlib, we find that he drew much of his knowledge of husbandry from the Americas, a fact about his archives that historians have mostly overlooked. Hartlib collected petitions, for example, that proposed social reforms based upon transporting sections of the population to the colonies. One John Lloyd argued for 'provision for the poor' in which he suggests that some 'may be sent over to the plantacion of [*blank*] in the West Indies, being one of the Charibbee Islands belonging to this Commonwealth'.[55] Further, the Hartlib Papers contain a multitude of letters making reference to individual colonies, especially those in the West Indies. There is, for example, a record of a 'Mr Dent' whose 'balsam was extracted' in the Bahamas.[56] In addition, the *Ephemerides* contain letters from the physician and agriculturalist Robert Child on the subject of Native Americans,[57] as well as Benjamin Worsley's extensive assessment of the merits of trade of the American plantations.[58]

We see the influence of the New World colonies not only in the *Ephemerides* but also in the published works of Hartlib and his friends. In *Samuel Hartlib his Legacie or An Enlargement of the Discourse of Husbandry used in Brabant and Flaunders; Wherein are bequeathed to the Common-Wealth of England more Outlandish and Domestick Experiments and Secrets in reference to Universall Husbandry*, many of the observations and experiments come not from Brabant and Flanders, but from the Americas. Under the heading 'Orchards', for example, Hartlib writes: 'I know, that in *Virginia* and *New-England*, that *Pines* and *Firres* and *Cedars*, do grow wonderfully thick in such *Moores* or *Swamps*, and being light wood, and easily wrought, they are continually used, while they last, for buildings.'[59]

Under the heading 'Dunging and Manuring Lands', Hartlib explains how 'in *New-England*, they fish their ground'. He notes that 'in the *North* parts of *New-England*, where the fisher-men live, they usually *fish* their ground with *Cods-heads*; which if they were in *England* would be better imployed'.[60] Similarly, New England is used as an example of the effective means of cultivating several commodities, including honey: '*Bees* thrive very much in *New-England*' and yet England is '*Deficient* in the ordering of them'.[61] Virginia is also held up as an example, especially in its cultivation of silkworms: 'It's a great *Deficiency* here in *England*, that we do not keepe *Silke-wormes* ... In *Virginia* also the *Silke-wormes* are found wild amongst the *Mulberry-woods*, & perhaps might be managed with great profit in those plantations if *Land* were not so scarce and deare.'[62]

Like *Samuel Hartlib his Legacie, The Commonwealth of Bees* contains a multitude of references to the Americas. Some of this information was sent to Hartlib from the physician and alchemist George Starkey who lived in Bermuda.[63]

Hartlib quotes Starkey's letters: 'In the Summer Islands, where I was born, I never yet saw one Bee, except those of a kind called Humble Bees.' Starkey lists a number of observations he has made in the colonies. 'In the Summer Islands I found, that in Rain-water kept in wooden Troughs ... there would gather a sediment of muddy matter to the bottome ... out of which would breede ... Wormes'.[64] Another 'very anomalous Generation, which I have noted, is of a sort of stinging fly [living] out of rotten Trees: these in the Summer Islands, I have observed out of the rotting Palmeto, and in New *England* I have seen the same in rotten Poplar and Birch'.[65]

The references to the Americas are too numerous to list in their entirety, but include the report that 'in the Summer Islands there is a sort of Spider, that is very large, and of admirable gay colours'.[66] Furthermore, 'there is a Berry ... *both in the Summer Islands and in New* England, *which makes me to conceive that it would also grow here*';[67] and 'again the Cochmeel, which is so rich a Commodity, cometh out of a fruit called the Indian Fig or prickled Pear, which as yet none of our Nation have attempted to make, which is the cause that die is so scarce, although *in all the Summer Islands the Tree bearing that fruit is most plentifull*'.[68] The 'prickled Pear' is later mentioned again: 'as to the Indian Fig ... and as to the transportation of it hither, here to grow, it is enough it growes there, being an English Plantation, or the fruit it self may ... be yearly brought over at rates reasonable enough'.[69]

The Compleat Husband-man contains reports from Virginia and New England on pine trees,[70] from Virginia on its vines,[71] on New England and Virginian wheat,[72] and how the people in the Bermudas fatten their cattle.[73] When analysing honey, Hartlib comments that because of the amount of sap emitted from trees, '*Bees* thrive very much in *New-England*'.[74] Here the knowledge that the American colonies can provide about agricultural improvement is especially important. This is because 'it's a great *Deficiency* here in *England*, that we do not keep *Silk-wormes* ... for to make silk ... In *Virginia* also the Silk-Wormes are found wilde amongst the *Mulberry woods*, and perhaps might be managed with great profit in those plantations.'[75]

Hartlib rails against what he calls the 'ignorance of the Husbandry of other places': 'I entreat them earnestly, not to think these things too low for them, and out of their callings; nay, I desire them to count nothing trivial in this kind, which may be profitable to their Countrey, and advance knowledge.'[76] Hartlib knew the types of plants that grew in New England:'Summer Wheat', he says, 'is sowen abundantly in *New-England*'[77] as well as 'Oats, white, black, naked, which in *New-England* serveth well for Oatmeal'.[78] The Bermudas 'have a peculiar way of fattening their Cattel, not used any where else that I know, which is with Green Fennel, that growth in that Island plentifully'.[79]

Hartlib explicitly compared the knowledge of plants and methods of husbandry in England with those in her colonies. He notes:

there are divers things in our Plantations worth the taking notice of, in Husbandry.
To passe by the Southern Plantations, as *Barbadoes, Antego*, Saint *Croix-Christopher,*
Nevis, Montserate, where the commodities are only Cotton-wools, Sugars, Gingers,
Indicoes, which our cold climate will not produce; and also Tobacco which growth
also with us, about Norwich and elsewhere.[80]

Despite their abundance, Hartlib argues, the English are ignorant of the qualities of the colonies' plants: 'So we are ignorant what their *Far* or fine Bread
Corn was, what their *Lupine, Spury* and an hundred of this kind ... infinite are
Plants which we have ... and dayly new Plants are discovered, usefull Husbandry,
Mechanicks and Physick.'[81] In addition to the ignorance of knowledge of the
colonial plants, Hartlib argues that the English also know very little about the
plants they grow at home:

> Though many hundred Plants do grow amongst us; yet but few of them are used
> Physically ... And truly in my opinion it is a great fault that we so much admire those
> things, that are far fetched and dear bought ... and corrupted by long voyages by sea,
> ... and do neglect those medicines which God hath given us here at home.[82]

Hartlib continues the theme of the general ignorance of husbandry in *Samuel*
Hartlib his Legacie. He notes a variety of natural resources growing in the northern American plantations which, though unrecognized by the English, would be
extremely useful if grown in England:

> Why may not the *Silke-grasse* of *Virginia*, the *Salsaparilla, Sassafras, Rattlesnake-*
> *weed* (which is an excellent Cordial) be beneficial to us, as also their *Cedars, Pines,*
> *Plum-trees, Cherries, great Strawberries*, and their *Locusts* ... be useful to us? So for
> *New-England*, why should we thinke, that the *Indian Corne*, the Marsh-wheate, that
> excellent *Rie* ... *Pumpions, Squashes* ... *Musk-Mellons, Hurtleberries, wilke Hemp, Fir*
> *&C*, of those parts, are altogether uselesse for us? ... But some will object, that they
> will not grow here with us, for your forefathers never used them. To these I reply, and
> ask them, how they know? Have they tried? ... So many things are found out by us,
> altogether unknowne to them, and some things will be left for our posterities: for
> example ... the *New World* and the wonders there, which not withstanding are but a
> few 1000 years standing: I say 20. *Ingenuitiies* have been found even in our dayes.[83]

The fact that the English know so little, as Hartlib made clear in the above quotation, could be solved by pursuing agricultural experiments:

> I desire Ingenuous Gentlmen and Merchants, who travel beyond Sea, to take \notice
> of the Husbandry of those parts (*viz.*) what grains they sow? At what time and sea-
> sons? On what lands? How they plough their lands? How they dung and improve
> them? ... also what books are written of Husbandry, and such like? And I intreat them
> earnestly ... to inform me in some trivial and ordinary things done beyond sea.'[84]

As Hartlib summarized his point:

the *Deficiency* is, that *Gentlemen* try so few Experiments for the advance of this honest and Laborious Calling … I hope in time there will be erected a *Colledge of Experiments*, not only for this, but also all other *Mechanicall Arts*.[85]

What do all these references to knowledge gleaned in the New World colonies amount to? The analytical difficulty here is that the information is scattered throughout the Hartlib Papers, such that they do not lend themselves to a systematic study of any particular text. We are limited, therefore, in that we cannot really engage in a *systematic* study of the Hartlib Circle's interests in colonization. We can, however, conclude from the evidence that England's Atlantic colonies were a formative context for the development of the Hartlib Circle's natural philosophy.

The second way in which the Atlantic colonies played a vital role in the Hartlib Circle's natural philosophy was by providing a space for agricultural experimentation. Colonies were well suited to the pursuit of natural philosophy in both their purpose and epistemology; they were controlled environments in which useful knowledge was extracted in order to improve the world.

In order to make husbandry a productive and useful endeavour, information about crops had to be ascertained through observation and rigorous testing. Which are the best crops to grow in particular areas? How do the climate and soil affect different crops? What yield can be expected? What tools and machinery are best for which crops? To engage in husbandry for the good of the Commonwealth necessitated an experimental enterprise.

Hartlib went to great efforts to explain the advantages of experimenting with methods of husbandry in plantations. In his *Cornucopia: A Miscellanium of Lucriferous and most Fructiferous Experiments, Observations, and Discoveries … to be Really Demonstrated and Communicated in all Sincerity*,[86] for example, he emphasizes that its experiments are designed for ' the good of the publick'.[87] They include, for example, 'An excellent experiment to make trees bear much and exceeding good fruit',[88] and 'another great experiment in ordinary grounds, without muck, which (by a new invention) five acres thereof have this last year afforded above 200 lib benefit in one acre'.[89]

The experimental nature of planting can be seen in two major characteristics of husbandry experiments. The first is that they had to be repeatable by fellow planters throughout England and her colonies. Referring to 'a cleer demonstration … to make exceeding rich muck in abundance, 1,000 loads and upwards' and an 'infallible experiment, to be confirmed by principles of sound reason … to draw forth the earth to her uttermost fruitfulnesse, and to moisten, fatten and fertilize sandy, dry and hilly grounds', Hartlib makes clear his intention that 'these two last may be performed accordingly throughout *England*'.[90] Hartlib's tracts such as this were in many senses advice books, so the advice they provided had to be reliable such that the experiments could be repeated.

In order for experiments to be replicated, their details needed to be recorded precisely. One example of such a project was the tract that Hartlib published on silkworms. Hartlib meticulously describes the type of soil '*fit for the* Mulberry-seeds, *how the same is to be ordered, and in what sort the seeds are to be sowed therein*'.[91] He goes on to describe the way that '*When the Plants that are sprung up of the Seeds,* [they] *are to be removed*'.[92] Hartlib's description of these steps continues for several pages and extends to the details of '*when and how to make fit rooms for the worms to work their bottoms of silk in, and in what sort the said bottoms are to be used*'.[93]

It was not only silkworms that were the subject of Hartlib's husbandry experiments. *The Reformed Commonwealth of Bees* to which 'The Reformed Virginian Silkworm' was an addendum, displayed the fundamental elements of a series of experiments. The *Commonwealth of Bees* consisted of a collection of letters detailing various methods for keeping bees and constructing the best beehive.[94] Many of the letters set out a particular method and experiment. One such was *An Experiment of the Generation of Bees*:[95]

> Take a Calf, or rather a Sturk ... of a year old ... bury it eight or ten dayes, till it begin to putrifie and corrupt; then take it forth of the earth ... After a while when [the maggots] begin to have wings, the whole putrified carcasse would be carried to a place to prepare, where the Hives stand ready, to which, being perfumed with Honey and sweet hearbs, the Maggots will resort.[96]

Other experiments were less gruesome, but they all set out specific instructions. A second characteristic of many experiments was their use of instruments. Hartlib introduced a letter detailing 'that pleasant and profitable Invention of a Transparent Bee-hive', written by Christopher Wren.[97] Wren describes the beehive and, using the language of experiment, he desires 'you farther light in this business ... from other mens Observations, that have tried the like Experiment'.[98] Hartlib then put forward another set of '*Observations concerning the swarming of Bees together with a short Description of a Bee-hive made of Glasse*'.[99] He also presented a diagram of an octagonal beehive which details its construction in a correlating key.[100]

The experimental nature of the Hartlib Circle's natural philosophy alerts us to a complicated epistemological issue. Recent scholarship in the field which Lorraine Daston calls 'historical epistemology'[101] has drawn our attention to the emergence of the modern fact as a distinctive development of the seventeenth century. Before that time, as Peter Dear has shown, the most important type of information was common experience because, in the Aristotelian tradition, it pointed towards knowledge which was universal. For a variety of reasons arising out of Francis Bacon's critique of Aristotelianism, however, the seventeenth century witnessed an epistemological shift in which natural philosophers began to privilege individual or unique events above Aristotelian commonplaces.[102] In short, what emerged from this process was the idea of facts as 'nugget[s] of experience detached from theory', to quote Lorraine Daston.[103] As Mary Poovey puts it, these modern facts were at once 'observed particulars *and* evidence of some theory'.[104]

What is somewhat intriguing about the Hartlib Circle is that their natural philosophy was based upon experiments, and the crucial issue was the usefulness of colonies in a project of universal reformation. The evidence presented in this chapter is full of examples of excitement about the multitude of experiments that can be carried out in colonies (silk and beehives to name just two), but this is not often accompanied by concerns about the veracity of knowledge these colonial experiments produce. Colonies do not appear as epistemologically problematic for most of the members of the Hartlib Circle in the 1630s and 1640s.

The one place where Hartlib expresses an interest in the reliability of knowledge is in his descriptions of the kinds of men who would make good correspondents in his proposed College of Husbandry. Hartlib suggested that 'there may be a *Colledge* or *School* of all the sorts and parts of *Good-Husbandry* erected … so the knowledge and practices [become] more universal, and men may have more sweet invitations and stronger allurements, to seek the knowledge of this deep and excellent *Mystery*; and practice it to the advancement of a more *general* and *Publique good*'.[105]

Hartlib's college aimed to build a community of men whose knowledge could be relied upon. We need only read Hartlib's careful description of the qualities of men who would be allowed into the college to realize that his intention was to use such a college as an institutional means of verifying knowledge collected in plantations. At the conclusion of *Cornucopia*, for example, Hartlib wishes that:

> further by men of worth, knowledge & experience, be respectively intelligenced, instructed and advertised of the manner and condition of traveling into what part soever … Whosoever shall have relation to *Virginia,* the *Barbadoes,* New *England,* or any other Countrie inhabited with English, or shall have cause to send into any of those places, or would inhabit, or transplant himself into those parts, he may have all intelligence and expedients, with as much conveniency as may be.[106]

Men sending information would have to be sons or kinsmen of 'a *Person* of quality'.[107] Alternatively, 'into this *Colledge* also any man may *enter* himselfe as a *free-man,* or *Friend* to, and *Member* of the *Society* upon the following conditions'.[108] These included the requirement that 'he must bring with him some *skill,* at least *ingenuity*; and testifie himself to be a *Well-wisher* to the *profession* and *professors of Good-Husbandry*'.[109] The sense of community would be built upon the fact that the members 'must always be in *Commons* at the *Hall* of the said

Society.[110] Moreover the men shall promise 'under hand & seale' to be 'a faithful *Seeker* of the *Advancement* of the *Mystery* and *Society*'.[111] Above all, one must not have a wife.[112]

We need only to observe how Hartlib described the men who gave him information to realize that knowledge from plantations, both foreign and domestic, could be trusted because it was produced by gentlemen, that is, by men who shared education, wisdom and the same intellectual goals. In *Some Proposalls Towards the Advancement of Learning* (1653), Hartlib outlined specific qualities necessary for natural philosophers, most of which were the product of being an efficient husbandman. 'The proper qualifications ... for the service of the publick' included 'Sciences which fit men with necessary knowledge for such works' and 'are comprehended under these Arts as we conceive: 1. The Art of Husbandrie, and what belongs thereunto; 2. The Art of Navigation and what belongs thereunto; 3. The Art of Mineralls, and what belongs thereunto; The Art of Surveying Lands ... The Art of Architecture ... the Art of Painting ... The Arts of working in Metalls.'[113] Most of these were vital skills for planters and therefore vital for natural philosophers.

Hartlib's concerns here aside, the majority of the Hartlib Circle's work points towards a contrast with Boyle on the issue of epistemological reliability. Although Boyle was in correspondence with the Hartlib Circle, it is useful here to distinguish between the Hartlib Circle's writings and correspondence during the decades leading up to the Restoration, and Boyle's work in Oxford in the 1660s. As Steven Shapin and Simon Schaffer have shown, the advent of the Restoration and the establishment of the Royal Society generated a new approach to the problem of knowledge. It was in this context that Boyle carefully conceptualized the 'fact' and the protocols of experimentation which made the production of modern facts possible and legitimate.[114] Boyle, unlike members of the Hartlib Circle, was explicitly concerned with epistemological questions. This is not to say that the question of knowledge did not play a role in the Hartlib Circle's intellectual arena, but rather that Hartlib and his colleagues did not problematize epistemological questions in the way that Boyle did.

This contrast between the Hartlib Circle and Boyle, which will become clear in the next chapter, enables us to perceive an important historical development in the conception of colonies. For the Hartlib Circle, colonies were an ideal space for the reform of knowledge and for agricultural experiments. Boyle's analysis of colonies, however, was more complex. As we will see, by the time of the Restoration, Boyle was arguing that the recreation of Adamic empire and the pursuit of English Protestant colonization were part of the same enterprise. Moreover, for Boyle, the issue of the reliability of colonial knowledge was central.

Natural Histories, Political Arithmetic and Ireland

Thus far, this chapter has examined the ways in which colonial husbandry and planting in the Americas underpinned the Hartlib Circle's natural philosophy. Let us move back across the Atlantic to Ireland where the Hartlib Circle's husbandry helped produce a new and quantitative kind of writing about colonies. This was political economy.

The ideal of recovering man's dominion over nature is the correct lens through which to view Petty, Worsley and the Boate brothers' actions in Ireland. The group were recovering dominion over Ireland not just through the act of colonizing but through producing and administering scientific knowledge.

The dedicatory epistle of the second edition of Boate's *Natural History* places the Irish project firmly in the context of the restoration of man's dominion over the earth, and refers to 'the time of *the Restauration of all things*'.[115] At this moment, 'the Intellectuall Cabinets of nature are opened, and the effects thereof discovered, more fully to us, than to former Ages'.[116] So the natural history of Ireland was part of a project in which 'knowledge of the natural History of each nation' was to be 'advanced & perfected'.[117]

The scholarship on the Hartlib Circle's interest in Ireland has tended to view the country as a blank slate for the application of Hartlibian natural philosophy. In her essay 'Natural History and Historical Nature: the Project for a Natural History of Ireland', Patricia Couglan rightly cites the Baconian vision of the 'enlarging of the bounds of Humane Empire, to the Effecting of all Things possible', as the intellectual context for what Petty did in Ireland.[118] She assumes, however, that the Baconian ideal was simply applied to Ireland: 'Bacon's ideal of the transformation by science of the material conditions of life was being *applied* by this group to Ireland.'[119]

Although it is certainly true that the ideal of recovering dominion over nature shaped the treatment of Ireland, the relationship between the new philosophy and Ireland was not simply a one-way street. Let us explore the possibility that the traffic of ideas and influence also moved in the opposite direction. The act of colonizing, I suggest, produced a new type of natural philosophy.

As the English colonization of Ireland intensified so did the task of administering the captured lands. Making colonies economically useful to England required that detailed natural knowledge be gathered. How profitable, for example, was a certain section of land? What value should be placed upon particular crops? Gerard Boate began compiling his natural history of Ireland in 1645, drawing upon information gleaned by his younger sibling Arnold, a physician in Ireland, from Sir William Parsons the Surveyor-General of Ireland, as well as from various other contacts.[120] After Boate died in 1653 the physician and agriculturalist Robert Child was placed in charge. Child had travelled to Ireland in 1651, and had corresponded regularly with Hartlib, who published his letters on agriculture in his *Legacie of Husbandry*.

In the dedicatory epistle to Boate's *Ireland's Natural History*, John Dury was explicit that natural histories were the kind of knowledge needed in order to render colonies profitable:

> For whether we reflect upon that first settlement of a Plantation, to prosper it, or upon the wealth of a Nation that is planted, to increase it, this is the Head spring of all native Commerce and Trading ... I know nothing more usefull, than to have the knowledge of the Naturall History of each Nation advanced & perfected.[121]

Coughlan interprets the natural history of the Boate brothers as an instance of a 'gaze that objectifies and examines'. 'The Irish', she writes, 'are objectified, along with their landscape, in the polemic of their representation as savage and uncivil.'[122] This is certainly an apt analysis. Without putting too fine a point on it, however, we should not assume that this gaze pre-existed and was *applied* to Ireland. Rather, the exigencies of administering a large quantity of Irish land in such a way that that its resources could be made useful produced the need for quantitative knowledge. Put simply, the act of producing quantitative knowledge of Ireland – the 'gaze that objectifies and examines' – arose from the nature of colonization.

The Hartlib Circle's desire to use knowledge gleaned in plantations for the profit of England created the need for a type of natural philosophy that could *claim* to be exact, quantifiable and objective. Petty's system of political arithmetic aspired to these ideals, arguing that the act of quantification and the reduction of goods to numbers was a truly objective and disinterested practice. Petty needed to *claim* to be producing a disinterested survey of Ireland. As Mary Poovey and others have shown, while acting as physician to the lieutenant-general of Cromwell's army, Petty acquired 19,000 thousand acres of Irish land, a number which increased to 50,000 by 1685, when his annual income amounted to £6,700.[123]

Petty conducted the Down Survey in 1655–6 while Worsley was the Surveyor-General of forfeited estates in Ireland. The Down Survey assigned value to the land, thus making both the settlement of debts and the creation of future profit possible. Petty studied the '*Territory, People, Superlucration*, and *Defencibleness* of both *Dominions*, and in some measure of their Trade, so far as we had occasion to mention *Ships, Shipping*, and nearness to *Ports*'.[124] The act of trading brought with it the need to quantify goods and profits. Petty calculated:

> the value of *Sugar, Indico, Tobacco, Cotton*, and *Caccao*, brought form the Southward parts of *America*, Six hundred thousand pounds. The value of the *Fish, Pipe-staves, Masts, Bever &c*, brought from *New-England*, and the Northern parts of *America*, Two Hundred Thousand pounds.[125]

Petty needed to quantify exactly how many people could engage in trade and what benefits they could yield. He concluded that there were 780,000 people 'fit for trade',[126] and he later established that Ireland 'lieth Commodiously for the Trade of the new *American* world; which we see every day to Grow and Flourish'.[127] Specifically, 'it lyeth well for sending Butter, Cheese, Beef, Fish to their proper Markets, which are to the Southward, and the Plantations of *America*'.[128] Towards the end of *Political Arithmatick*, he makes valuations, that 'all *Cloaths, Houshold staff, &c.*, carried into *America*' was worth £600,000,[129] and 'The value of *Sugar, Indico, Toabacco, Cotton and Caccao*, brought from the Southward parts of *America* Six hundred thousand pounds'.[130]

This kind of calculation was also needed to assess the relative value of husbandmen when compared with seamen. 'There be in *England, Scotland, Ireland* and the Kings other Territories above Forty Thousand Seamen; in *France* not above a quarter so many; but one seaman earneth as much as three common Husbandmen; wherefore this difference in Seamen, addeth to the account of the King of *England's* Subjects, is an advantage equivalent to Sixty Thousand Husbandmen.'[131] Furthermore, he wrote, 'may not the *English* in the *America* Plantations (who Plant *Tobacco, Sugar, &c*) computer what Land will serve their turn, and then contract their Habitations to that proportion, both for quantity and quality'.[132]

As the organization and control of colonization increased, so the discourse in which it was understood became more quantitative and meticulous. It needed to, in order to render colonial knowledge useful. Natural philosophy became more theorized, experimental and regulated as a result of the exigencies of effective planting. Political arithmetic produced the kind of knowledge which colonization demanded: quantifiable, and encased in a claim of epistemological reliability. Nature produces exact knowledge that can be expressed in numbers, weights and measures. As Petty states in his Preface, his aim is to 'consider only such Causes, as have visible Foundations in Nature'.[133] Petty describes his method thus: 'I have taken the course (as a Specimen of the Political Arithmetick I have long aimed at) to express my self in Terms of *Number, Weight,* or *Measure*'.[134] Our analysis of Petty's work suggests that we should modify the conventional view of the relationship between the Hartlib Circle and Ireland. Of course Ireland was a space for the Hartlib Circle to apply their natural philosophy of resurrecting the original empire of man. Conversely, however, Irish colonization helped to shape the study of nature. The relationship between natural philosophy and colonization, therefore, was reciprocal. Petty's calculative, theoretical discipline of political arithmetic, and the Hartlib Circle's increasing attention to the rigours of experiment suggest that Irish colonization played an important role in the foundations of experimental natural philosophy.

By instigating an avid interest in the American colonies as sources of natural philosophical knowledge and fostering the close relationship between the

scientific experiment and the act of planting, the Hartlib Circle helped introduce colonization into the discourse of epistemic empire. Establishing colonies
became a way of recovering Adam's dominion over nature.

Placing the Hartlib Circle in an imperial and colonial context enables us to
understand the historical trajectory of the concept of empire. In the previous
chapter we saw that, while Francis Bacon emphasized the application of knowledge and its responsible use for the good of the Commonwealth, he did not
conceive a role for the Atlantic colonies in rebuilding man's plenary empire.
By contrast, the Hartlib Circle's work brought colonization into the picture,
and with it the emphasis upon the agrarian dimension of the renewal of man's
original empire. Long excluded from histories of the British Empire, the Hartlib
Circle and their natural philosophy should now be written into the story.

3 ROBERT BOYLE'S PROTESTANT COLONIAL PROJECT

The Opportunity I had of being one of the Committee or Directors of the English *East-India* Company, (whereto the desire of Knowledge, not Profit, drew me).

Robert Boyle, *Experimenta et Observationes Physicae* (1691)[1]

Buried in the chapter 'Containing Various Observations about Diamonds', within his diverse collection of scientific experiments entitled *Experimenta et Observationes Physicae* (1691), is Robert Boyle's most succinct statement about his interest in English maritime trade. His motivation was knowledge. Given Boyle's desire for information collected by employees of the English East India Company, this chapter attempts to answer the following question. Was there a connection between Boyle's interest in English commerce and colonies, and the rest of his work?

The idea that the New World and its colonies might yield useful knowledge for natural philosophers was common among members of the Hartlib Circle. I suggest, however, that Boyle gave this idea a detailed theological underpinning found neither in the work of Francis Bacon nor Boyle's Hartlibian colleagues. In his defence of experimental philosophy, Boyle drew a theological connection between natural philosophers' pursuit of man's original empire over the world, and English trade and colonization, which he argued were the means of fulfilling God's command to man to enjoy the fruitfulness of the earth. In short, Boyle gave a theological framework to the idea that the recovery of Adamic empire and the pursuit of an English Protestant colonial empire were part of the same enterprise.

This chapter attempts to bring together an understanding of various aspects of Boyle's work: his theology, his natural philosophy and his colonial interests and activities in the Council for Foreign Plantations, the New England Company, the East India Company, as well as his extensive correspondence with the New World. I will suggest that identifying Boyle's Adamic ideal of restoring the '*Empire of Man over inferior Creatures*'[2] enables us to perceive a continuity between these various dimensions of Boyle's work, among which there is sometimes a disjunction in the historiography.

In the past three decades, two developments in Boyle studies have transformed the field. The first, broadly considered, is a contextual reading of Boyle: an attempt to understand what Boyle was 'doing' with his natural philosophy and why. This line of enquiry has improved our understanding of Boyle and his colleagues' 'social vision', to use James Jacob's term. 'The latitudinarians of the Royal Society', Jacob argued, 'were equally devoted to the same social vision, the joint purposes of science, trade, empire and reformation.'[3] This chapter builds upon Jacob's work, though I will take issue with his understanding of what Boyle meant by 'empire'.

Recently, Steven Shapin and Simon Schaffer have drawn attention to the importance of seventeenth-century debates about the relationship between knowledge and authority. By establishing an experimental method which could produce reliable matters of fact, Boyle aimed to provide a solution to the problem of knowledge and political order.[4] The final section of the chapter attempts to build upon Shapin and Schaffer's work by examining the interviews which Boyle conducted with travellers and recorded in his *Work-Diaries*, focusing upon the way he dealt with the epistemological issue of the veracity of information collected on the colonial periphery. The evidence suggests that Boyle used men's associations with England's colonies as a means of establishing their gentlemanly reliability.

The second recent development in Boyle scholarship, inspired primarily by the work of Michael Hunter, is the recognition of Boyle's complexity as a thinker. Hunter's detailed investigation into Boyle's 'scrupulosity', for example, has shown how his fastidious nature, grounded in his deep religious convictions, manifested itself in the assiduous manner in which he pursued natural philosophy.[5] One of Hunter's insights is the 'clear parallel between the indefatigability that Boyle displayed in the laboratory and the assiduity with which he sought to salve his conscience through casuistry'.[6]

Hunter argues that much of the scholarship in the tradition of the sociology of science has neglected to examine carefully Boyle's intricate religious beliefs: 'Boyle is made to seem too secular minded in his motivation, too concerned about the political threat from Hobbes and the legacy of the Civil War, when in fact his concerns were dominated by religion to an extent which may seem implausible to late twentieth century commentators.'[7] I do not think, however, that there needs to be a tension between the two readings of Boyle. It is possible to appreciate both Boyle's deep religiosity as well as the political and epistemological contexts in which he wrote. The chapter will attempt to advance this view by showing how one aspect of Boyle's theology – his conviction about restoring man's dominion over nature – underlay his interest in English colonization. Colonies and trade, Boyle believed, were means of extracting the riches of the earth intended for man. As such, they contributed to making the world fruitful and restoring man to his original empire over nature. It is my contention that Boyle conceptualized English colonization and the restoration of man's dominion over nature as part of the same enterprise.

Boyle's Theology

Religious convictions characterized Boyle's life.[8] The fact that he framed his vision of natural philosophy in theological terms, therefore, should not surprise us. Born in 1627, Boyle was the son of the youngest child of Richard Boyle (1566–1643), the first Earl of Cork and Lord High Treasurer of Ireland. At the age of thirteen, Boyle travelled abroad to France, Italy and Switzerland. During this trip he experienced a religious epiphany precipitated by a thunderstorm so terrifying that it engendered the thought of the day of judgement.

After his return to England from the continent in 1644, Boyle settled on the estate of Stalbridge in Dorset and became involved in the short-lived Invisible College which pledged to use science for the Puritan ends of reforming and improving society.[9] A number of members of the Invisible College were closely associated with the Anglo-Irish circle centred upon Boyle's sister Lady Ranelagh.[10] These included the brothers Arnold and Gerard Boate, as well as Benjamin Worsley.

Upon the death of Gerard Boate, Hartlib asked Boyle to complete *Ireland's Natural History*, but Boyle declined.[11] We do not know why, but one could hazard a guess that Boyle's Irish interests, and the wealth which stemmed from them, sometimes drew him undesired attention. Boyle was a wealthy and eventually a famous man, and he was frequently asked to become a patron or to serve in various offices, some of which he simply had to decline.[12] Furthermore, his family connections in Ireland were well known. Both his brother Roger, Lord Broghill, and his sister Lady Ranelagh played leading roles in the political and military affairs of the Protestant community in Ireland.[13] It should not really surprise us, therefore, that Boyle gives very little attention to Ireland in the corpus of his published work. He was no doubt very conscious of the origins of his wealth and his silence about Ireland will become apparent throughout this chapter.

Boyle's intellectual associations developed when he moved to Oxford in 1654 and established a laboratory with Robert Hooke.[14] Whilst in Oxford, Boyle corresponded with the Governor of Connecticut, John Winthrop, the missionary John Eliot and the astronomer Seth Ward. Oxford intellectual life introduced Boyle to the study of oriental languages including Hebrew, Aramaic and Syriac, all of which he mastered. His passion for philology was reflected in his correspondence about a universal language, and in his support for the translation of the Bible into Irish and Algonquian. Boyle's linguistic and missionary interests were intimately connected: the dissemination of knowledge through proselytizing was one embodiment of the ideals of reformation, epistemic empire and colonization.

In 1649, Parliament passed the Ordinance for the Advancement of Civilization and Christianity among the Indians, which created the Society for the Propagation of the Gospel in New England. When the Society was incorporated as the New England Company in 1661, Boyle became its first governor.[15] He took a careful interest in the Society's missionary and charity work which included 'educating, Cloathing, civilisinge and instructinge the poor Natives' and moreover, 'puttinge an industrious People into a way of Trade and commerce ... but more especially to endeavour the good Salvation of their immortal Soules, and the publishing the most glorious Gospell of Christ amongst them'.[16]

Boyle became a member of the governing Council of the Royal Society of London and attended the first meeting in November 1660. Together with a number of other Fellows of the Society, including Henry Oldenburg, Boyle was also active in the Council for Foreign Plantations, to which he was commissioned in December 1660 and served until 1664. In this role, Boyle's responsibility was to 'oversee the administration of the English Colonies in the West Indies and North America', part of which involved ensuring the Navigation Acts were enforced and that the King's Commissioners to the United Colonies of New England carried out their work properly.[17]

In the Royal Society, the new philosophy's attack upon scholasticism found an institutional home, and Boyle became one of the leading protagonists with *The Sceptical Chymist* (1661), *The Origin of Forms and Qualities* (1666), *Experiments and Considerations Touching Colours* (1664) and *New Experiments and Observations Touching Cold* (1665). In the latter two works Boyle aimed to vindicate a corpuscular explanation of his findings over a scholastic reliance upon the concepts of forms and qualities. Boyle's experimentalism in his first scientific book, *New Experiments Touching the Spring of the Air* (1660), which used the air pump he developed with Robert Hooke in order to prove, contrary to the belief of the Scholastics, the possibility of vacuums in nature.

During the 1670s, Boyle became involved in a debate with the Cambridge Platonist Henry More about the intervention of God in the natural world. More misused Boyle's findings in *New Experiments Physico-Mechanical Touching the Spring of the Air* in an attempt to argue for a 'spirit of nature' which controlled the natural world. Boyle, however, was vehemently opposed to any concept of a spirit of nature because, as J. E. McGuire has shown, he adhered to a nominalist ontology in which God's power in nature was absolute because particulars in nature are unrelated and are 'denied the power to cause change in and of themselves. God's Will, therefore, is the only causally efficacious agency in nature.'[18]

The last decade of Boyle's life before his death in 1692 was marked by a deeper investigation into the theological issue of the relationship between God and the natural world.[19] In this context he published *A Discourse of Things above Reason* (1681), *A Free Enquiry into the Vulgarly Receiv'd Notion of Nature* (1686), *A*

Disquisition about the Final Causes of Natural Things (1688) and *The Christian Virtuoso* (1690). An exploration of the idea of Adamic empire in Boyle's theological writing offers us an excellent point of entry into his natural philosophy.

> Tis recorded in the Book of Genesis, the Design of God in making man, was, that men should Subdue the / Earth (as vast a Globe as 'tis) and have dominion over the Fish of the Sea, and over the Fowle of the Air, and over the Cattle, and over all the Earth, and (to speak Summarily) over every living thing that moveth upon the Earth.[20]

Boyle's use of the idea of man's original empire over nature in this essay illustrates the way it served as a tool to legitimate experimental philosophy. One of Boyle's chief means of defending the new philosophy against its various detractors was to argue that it was uniquely suited to accomplish man's original imperial project; it was the only philosophy that enabled man to exploit the world as God intended. The critics of experimental philosophy included scholastics, Hobbesians such as Henry Stubbe, and also men with whom Boyle disagreed about specific concepts, such as the Cambridge Platonists and their idea of a spirit of nature. In addition, there was a general sense that the Restoration era was plagued by a culture of atheism. As Hunter has shown, it is difficult to pinpoint any 'atheist' in particular, and indeed Boyle rarely mentioned his intellectual sparring partners by name, but we should understand Boyle's anxiety about atheism as a response to a fashionable Restoration culture of wit and a scepticism of many received ideas, one of which, Boyle feared, was belief in God.[21]

Although not naming his critic, Boyle tells us in *Final Causes* that 'a late Ingenious Author did causlessly reflect upon me, for having / mention'd the Enlightning of the Earth, and the Service of Men, among the Ends of God, which he thought undiscoverable by us'.[22] Thus, Boyle continues:

> Many other Texts that show, how much God was pleas'd to intend mans welfare, and Dominion over many of his Fellow-creatures, might be here alledg'd. But I shall content my self to mention, what the Kingly Prophet sayes in the 8th Psalm, where speaking of Man to his Maker, he sayes; *Thou has made him, a little lower than the Angels, and hast crown'd him with Glory and Honour. Thou mad'st him to have dominion over the works of thine hands, and hast put all things under his Feet.*[23]

Far from being undiscoverable, one of the 'Ends of God', Boyle showed, was that man must assert his dominion over the earth which was made for him. Boyle refers to Adam in a number of his works, including *The Excellency of Theology, The Christian Virtuoso* and the *Usefulness of Experimental Natural Philosophy*,[24] the latter of which contains his most detailed discussion of the idea. A promotional tract, the *Usefulness of Experimental Natural Philosophy* explicated the advantages of experimental philosophy over scholasticism. Part 2 of the second section of the work is devoted to showing how experimental natural philosophy may be useful to the '*Empire of Man over inferior Creatures*'[25] as its subtitle reads.

Addressed to an imagined audience in the shape of a young friend named Pyrophilus,[26] the text argues that natural philosophy teaches man not only to 'Know Nature', but also 'in many Cases to Master and Command her'.[27] Such is the naturalist's power to 'work such changes among the Creatures, that if *Adam* were now alive, and should Survey that great Variety of Man's Productions, that is to be found in the shops of Artificers, the Laboratories of Chymists, and other well-furnished Magazines of Art, he would admire to see what a new world ... has been added to the Primitive Creatures by the Industry of His Posterity'.[28]

Objects which rely upon a naturalist's power to produce them are beyond the scope of the Scholasitcs, who did not concern themselves with man's 'productions', but rather only with the 'disguis'd repetitions' of existing authorial texts.[29] Emphasizing the benefits that accompany the study of the natural world, Boyle argued that understanding God's 'book' of Creation was necessary in order to comprehend His other book, the Bible: 'God has made some knowledg of his Created Book both conducive to the beliefe, and necessary to the Understanding, of his Written one.'[30]

In contrast with the study of the natural world and its 'great Variety of Man's Productions',[31] scholasticism offered no new knowledge: 'Most other Sciences, at least as they are wont to be taught, are so narrow and so circumscrib'd, that he who has read one of the best and recentest Systems of them, shall find little in the other Books publisht on those subjects, but disguis'd repetitions.'[32] Natural philosophy, on the other hand, deals with a subject matter that extends throughout the globe and the entire natural world. It is therefore uniquely able to grapple with the

> incredible variety of Vegetables, that the teeming Earth, impregnated by Gods *Producat Terra*, does in several Regions produce. Botanists have a pretty while since, reckon'd up near 6000 Subjects of the Vegetable Kingdom; since when, divers other not-described Plants have been observed by Herbarists; the chief of which will, I hope, be shortly communicated to the World.[33]

Boyle was acutely aware of the enormity of the natural world, and the fact that the New World was unknown to the ancients confirmed his belief that only experimental philosophy would enable man to comprehend its 'immense Fabrick' made by God's 'Almightinesse'[34] in the Creation. The 'Terrestrial Globe we Men inhabit, containes, besides all those celebrated Monarchies, those spacious (since detected) American Regions, that have been deservedly stiled *The New World*'.[35] The implication is that the New World of God's creation was beyond the ancients' – and scholastics' – comprehension. God ordained the whole world to serve man and only experimental philosophy could give him access to the New World.

Boyle reinforced this point later in the *Usefulness of Experimental Natural Philosophy* by referring to two discoveries which only natural philosophy had made possible. The first was the loadstone, which enabled the making of the compass, without knowledge of which 'those vast Regions of *America*, and all the Treasures of Gold, Silver, and precious Stones, and much more Precious Simples they send us, would have probably continued undetected.'[36] Then, referring to a second discovery, this time of the 'supposed antipathy between Salt-Peter and Brimstone', Boyle tells us that it gave rise 'to the invention of Gunpowder'. This has

> quite alter'd the condition of Martial Affairs over the World, both by Sea and Land. And certainly, true Natural Philosophy is so far from being a barren speculative Knowledge, that Physick, Husbandry, and very many Trades (as those Tanners, Dyers, Brewers, Founders, &c.) are but Corollaries or Applications of some few Theorems of it.[37]

It is natural philosophy, therefore, which makes possible the kind of study of nature that will enable man to understand how to make use of the world and its creatures. Thus, practical arts, trades and husbandry are best seen as 'Corollaries or Applications' of natural philosophy. The importance of husbandry to Boyle is well illustrated by his comments in the first essay of the second part the *Usefulness of Experimental Natural Philosophy*:

> I shall not dare to think my self a true Naturalist, till my skill can make my Garden yield better Herbs and Flowers, or my Orchard better Fruit, or my / Fields better Corn, or my Dairy better Cheese then theirs that are strangers to Physiology. And certainly, *Pyrophilus*, if we seriously intend to convince the distrustful World of the real usefulness of Natural Philosophy, we must take some such course, as that *Milesian Thales* did, who was by the Antients reckoned among the very first of their Naturalists.[38]

Thales of Miletus, Boyle tells us, 'made advantage enough of his skill' to engage in practical arts and husbandry. Thus, as Boyle explains in the next section:

> Me thinks, it should be a disparagement to a Philosopher, when he descends to consider Husbandry, not to be able, with all his Science, to improve the precepts of an Art, resulting from the lame and unlearned Observations and Practice of such illiterate Persons as Gardeners, Plow-men, and Milk-maids. And indeed, *Pyrophilus*, though it be but too evident, that the barren Philosophy, wont to be taught in the Schools, hath hitherto been found of very little use in humane Life; yet if the true Principles of that fertile Science were thorowly known, consider'd and apply'd, 'tis scarce imaginable, how universal and advantagious a change they would make in the World: For in Man's knowledge of the nature of the Creatures, does principally consist his Empire over them, (his Knowledge and his Power having generally the same limits.)[39]

Here Boyle's sentiment is familiar; it is Baconian in its aspiration to make the world useful to man through the correct use of knowledge and power, and Hartlibian in its emphasis upon man's task to reap the agrarian fruitfulness of the world. But, while Boyle's ideas about man's plenary dominion are consistent with the Baconian and Hartlibian tradition, I would argue that they find in his own writing a richer and more nuanced theological grounding. There are three aspects of man's original empire to which Boyle gives careful theological explication.

The first of these is the idea that husbandry and manual arts constituted part of man's plenary dominion over nature. While both the Hartlib Circle and Bacon lauded the importance of these practical arts, it was Boyle who, in an attempt to legitimate practical arts, explained their precise theological significance. They were the 'chieftest Instruments of Man's dominion over the Creatures'.[40] Boyle showed that husbandry formed part of the properly defined relationship between God, man and the book of nature because it enabled man to exploit the earth's resources:

> The Clouds water his Land, the Earth supports him and his Buildings, the Sea and winds convey him and his Floating-houses to the remotest parts of the World, and enable him to possesse every where almost all that Nature or Art has provided for him any where. The Earth produces him an innumerable multitude of Beasts to feed, cloath, and carrie him; of Flowers and Jewels to delight and adorne him; of Fruits, to sustaine and refresh him; of Stones and Timber.[41]

In order to exploit these resources, knowledge of practical arts was required because some natural phenomena and creatures could not be tamed merely by man's brute force. God

> makes most of the Creatures of the world visible to us, pay homage to him, and in some manner or degree do him service: God's liberality at once bestowing on him all those Creatures by endowing him with a Reason enabling him to make use of them; so that even those Creatures which he is not able to subdue by his Power, he is able to make serviceable to him by his Knowledge.[42]

It was the natural philosopher's task not only to discover new practical arts in order to reap the fruitfulness of the world, but also to revive those arts once known to Adam but lost in the passage of time. As Boyle explained in the *Usefulness of Experimental Natural Philosophy*:

> So tis the work of the Experimental philosopher, not onely to dive into the deep Recesses of Nature, and /thence fetch up her hidden Riches; but to recover to the use of Man those lost Inventions, that have been swallowed up by the Injuries of Time, and lain buried in Oblivion.[43]

Because God intended man to command an empire over the earth, and because natural philosophy alone would enable man to garner the skills and knowledge to fulfil this divine injunction, Boyle argued that to deter man from natural philosophy was also to deter him from pursuing the project for which God created him:

Having thus premised, *Pyrophilus*, that two of God's principal aimes in the Creation, were the manifestation of his own Glorious Attributes, and the Welfare of his noblest Visible Creature, Man; It will not be perhaps difficult for You to discerne, that those who labour to deterre men from sedulous Enquiries into Nature, do ... take a course which tends to defeat God of both those mention'd Ends.[44]

The second aspect of man's plenary empire to which Boyle gives careful analysis is that of the *nature* of the power this empire constitutes. The 'Empire of Man', Boyle wrote, 'as a Naturalist, over the Creatures, may perchance be to a Philosophical soul preserved by reason untainted with Vulgar Opinions, of a much more satisfactory kind of Power or Soveraignty, then that for which ambitious Mortals are wont so bloodily to contend'.[45] In the context of seventeenth-century debates about the origins and proper use of political power, Boyle was making a statement which contrasts political power, so easily abused and so bloody in its pursuit, with power over Creation, which by contrast is divinely sanctioned and draws man closer to God.

In the second section of the *Usefulness of Experimental Natural Philosophy*, Boyle made a second point about the nature of man's original power over the earth. Introducing the first essay, Boyle tells Pyrophilus that, having just demonstrated the 'Usefulnesse of Experimental Natural Philosophy to Physick', he will now proceed to show how it is 'also very serviceable to Husbandry, in all its subordinate parts, and to those other Professions that serve to provide Man with Food or Rayment ... the trades of Brewing, Baking, Fishing, Fowling, Building'.[46] These practical pursuits enable man to assert his dominion over the Creatures, whereas 'Physick' only enables man to control human bodies:

The Advancement of his Empire seems to consist more properly in the Inlargement of his Power over the other Creatures: Physick seeming rather to defend him against Revolts and Insurrections at home, than to increase his Power, and extend the Limits of his Empire abroad.[47]

In Boyle's metaphor, 'abroad' denotes that which is external to the human body rather than areas geographically abroad, but nonetheless his point remains: whereas 'physick' only enables man to master his own body, natural philosophy enables him to extend his empire over nature.

The third dimension of Boyle's Adamic theology is his characterization of man's dominion as a project to be undertaken by the Christian Virtuoso. For Boyle, the recovery of man's original empire was part of a programme for a model life as an experimental philosopher and as a Christian. Boyle argued that the Christian Virtuoso was one to whom 'God has vouchsafed more than ordinary degrees of knowledge, of the excellency and usefulness of his admirable works of creation and providence, to extol and celebrate the Divine author of them'.[48]

Thus, the Christian Virtuoso had a particular role in discovering and celebrating the natural world over which God had given him mastery:

> So many and so various creatures as we have mentioned, to be directly subjected by God to man's dominion, may be of great advantage to him, and of more than they have been, when once they shall be improved by sagacious and industrious virtuosi.[49]

Neither Bacon nor Boyle's fellow members of the Hartlib Circle explored the theological basis for their ideal of restoring man's original dominion over nature in this much depth. For Boyle, by contrast, the subject was one of his chief preoccupations in the *Usefulness of Experimental Natural Philosophy*. Instructive here is Hunter's observation that Boyle's aim was to 'achieve a reformed natural philosophy which would form part of a properly defined relationship between God and man'.[50] The restoration of man's prelapsarian dominion, I suggest, was one of the chief elements of this relationship. In order to exert dominion over the Creation, man must recover his knowledge of the natural world. Dominion and knowledge therefore go hand in hand. In Boyle's work the idea of man's empire became a detailed theory in which he explored the nature of empire as a form of power: its origins, theological legitimacy and finally its role as a pursuit for the model natural philosopher, the Christian Virtuoso.

Adamic Empire and English Colonies in the Atlantic

Boyle saw the restoration of man's dominion over the world as entirely consistent with an English project of colonization – and proselytizing – in the Atlantic. Two points are worth mentioning here. The first is the significance of English colonization of America to Boyle's work. We can see a strong continuity between Boyle's colonial interests and his theology of Adam's empire. The second is a broader point about the development of the idea of man's dominion over nature. Boyle strengthened the agrarian dimension of this tradition by making colonization central to the recovery of man's original empire. In Bacon's work, as we saw in Chapter 1, the chief concern with the restoration of man's dominion was with the recovery of knowledge, his prelapsarian omniscience of the natural world. Boyle, following the Hartlibian tradition, emphasized not only the recovery of natural knowledge from the New World, but also the injunction to reap the earth's fruitfulness. It is this agrarian dimension which established the connection between Adamic dominion and colonization. The point is not that Boyle was doing something particularly novel here, as he is chiefly Hartlibian in his agrarian ideas. Boyle did, however, provide the theological underpinning for this dimension of man's original empire.

As we saw in the above discussion of his theology, Boyle believed that the re-creation of man's dominion over the world would be partly accomplished through reaping the earth's fruitfulness through husbandry. Husbandry was particularly suited to being practised abroad because it was primarily in the New World that man came across plants which he had not previously encountered and which could benefit him. There is 'an almost incredible variety of Vegetables, that the teeming Earth, impregnated by Gods *Producat Terra*, does in several Regions produce'. [51] An example of such a vegetable, Boyle tells us, is the 'excellent kinde of Pepper, whose Shell tastes not unlike Cinamon, and smells so like Cloves ... having been lately gathered in *Jamaica* (where it abounds) and presented me by the inquisitive Commander of the English Forces there'. [52] Given this Edenic abundance, it is unsurprising that Boyle believed that man's empire over the creatures was properly pursued abroad.

The troubling question is: whose empire? Boyle frequently referred to 'man's' empire, and indeed many of Boyle's discussions about Adam's plenary dominion, as we have seen, were articulated in a universalist language in which the subject was mankind. In addition, some of the information about climate, plants and animals collected in the New World, upon which Boyle relied, was collected by foreign travellers. Moreover, Boyle often cites the natural histories of Guilelimus Piso and Jose d'Acosta. Yet, despite this universalist language and obvious reliance upon the adventures of foreign travellers, Boyle's interest in and involvement with the institutions of colonization and trade were resolutely English.

This ostensible tension in Boyle's work illuminates one of the central themes of the relationship between seventeenth-century natural philosophy and the early British Empire: the complex relationship between natural philosophers' universalist aims on the one hand, and their practices which built a very English empire on the other. As James Jacob and Charles Webster have shown, Boyle and his circle shared the belief that England must play the leading role in the Protestant reformation of the world. This reformation would be effected through the use of knowledge for the improvement of mankind. Importantly, this was conceived as a *universal* project. All facets of society were seen as subjects for reformation: morality, education, language as well as the creation of wealth. The emergent field of knowledge called political economy, pioneered at this historical moment by men like John Houghton, William Petty and Nicholas Barbon, was underpinned by the belief that an increase in England's wealth would improve the state of its people. Both trade and colonies increased England's prosperity and reaped the world's financial and agrarian fruitfulness. Moreover, colonies made it possible to enact a religious and social reformation upon indigenous peoples through proselytizing.

The universal reformation of the world that Boyle advocated was fundamentally about man's imperial task on earth: the restoration of his original empire over nature through industry. Boyle never explicitly addresses the tension between the universal and the particular; between an empire for mankind and an empire for England. This tension seems glaring to the historian, but it is indeed possible that it was a problem Boyle never perceived. This is because the underlying assumption in his work is that, as the leader of the Protestant world, England was therefore also in command of this imperial – and colonial – quest. Put simply, England had adopted the Protestant mantle for mankind.

Robert Iliffe has addressed the issue of the relationship between the universal and the national in the context of the early Royal Society. He argued that the Royal Society members were 'the proponents of a co-operative, international experimental philosophy' and yet they 'simultaneously argued that it was peculiarly English'.[53] Thus,

> it was a peculiarity of experimentalism of the Royal Society that its propagandists were internationalist at the same time as being hostile to many elements of foreign cultures. For them, the future of natural philosophy rested on persuading others of the unique philosophical fertility of English experimentalism, while their notions of Englishness were defined by, and actively co-produced along with, disparaging attitudes to foreign individuals and practices.[54]

A careful exploration of Boyle's attitudes to colonization enables us to add to Iliffe's analysis. The 'peculiarity of experimentalism', which Iliffe identifies, had a colonial dimension, manifesting itself not only in the Royal Society members' interest in gleaning knowledge from abroad but also in Boyle's attitudes toward English colonization. Let us examine the following extract from the *Usefulness of Experimental Natural Philosophy* in which Boyle discusses the sugar plantations in the English colony of Barbados:

> I might alleadg the Art of cultivating and gathering Sugar-Canes, and of ordering their Juice, as a recent Instance of the transplanting of Arts and Manifactures. For, as I am inform'd by very credible Relations, there are not yet very many years efflux'd, since, in our memory, a Forreigner accidentally bringing some Sugar-Canes, as Rarities, from *Brasil* into *Europe*, and happening to touch at the *Barbadoes*, an English Planter that was Curios, obtain'd from him a few of them, together with some Hints of the way of cultivating and useing them. Which, by the Curiosity and Industry of the *English* Colony there, were in a short time so well improv'd, that that small Island became, and is still, the chief Storehouse that furnishes, not onely *England*, but *Europe* with Sugars. And this Instance I the rather mention, because it is also a very notable one, to shew, how many Hands, the Introduction of one Physico-Mechanical Art may set on work.[55]

What impressed Boyle about Barbados was the way that sugar cultivation had improved the colony and rendered it fruitful enough to be the 'chief Storehouse' for both England and Europe. Here, then, Physico-Mechanical Arts, which Boyle sees as instruments of man's dominion over the earth, have been put to work in a colony. Yet the chief beneficiary here is England rather than mankind.

The fact that Barbados was able to supply not only England but also Europe with sugar increased its prosperity.

The Navigation Acts of 1651 and 1660 were intended to augment English trade and limit the Dutch. Boyle was well aware of the benefits which trade and commerce brought to the colonies, particularly because of his membership of the Council for Foreign Plantations from 1660 to 1664. Let us examine what Boyle writes about trade in relation to Barbados as he continues his discussion of the colony in the second section of the *Usefulness of Experimental Natural Philosophy*:

> And that you may see how Luciferous in that place this so recent Art of making Sugar is, not onely to private men, but to the publick; I shall adde, that by divers intelligent and sober persons interested in the *Barbadoes* ... I have been inform'd, That there is, one Year with another, from that little Island, which is reckon'd to be short of thirty Miles in length ... shipt off for *England* ... ten thousand Tun of Sugar ... which amounts to twenty Millions of Pounds of that Commodity.[56]

Boyle's general and very Hartlibian point is that husbandry and 'Physico-Mechanical Arts' enable man to transform the uncultivated soil in colonies such that they yield raw, useful materials. Boyle's intention in the particular essay from which these quotations derive is to show how

> the Naturalist may increase the Power and Goods of Mankind upon the account of Trades, not onely by meliorating those that are already found out, but by introducing new ones, partly such as are in an absolute sense *newly / invented*, and partly such as are *unknown* in those places, into which he brings them into request.[57]

The colony of Barbados, then, was just such a place in which a mechanical art could be used to exploit the agrarian potential of the land. With his references to exploiting Barbados, Boyle adapted and modified the Adamic language of agrarian fruitfulness to justify the building of a very *English* colonial empire, replacing man with England as the subject of the divine historical plan. Thus, Boyle used the universal language of mankind's dominion as a general description of processes that in reality benefited English colonies. Several pages preceding his aphorism about Barbados, for example, Boyle introduced his general discussion about the transplantation of arts and manufactures in a universalist language, referring to trades that 'minister to the necessities of Mankind',[58] yet it is clear in his discussion of Barbados that the real intended beneficiaries of this transplantation of arts and manufactures are England and her colonies.

The fifth aphorism in the second part of the *Christian Virtuoso* is a good illustration of Boyle's appropriation of the Adamic language of man's plenary dominion to describe the benefits of English colonization, this time in Virginia as well as Barbados:

That so many and so various creatures as we have mentioned, to be directly subjected by God to man's dominion, may be of great advantage to him, and of more than they have been, when once they shall be improved by sagacious and industrious virtuosi, may appear very probable, if we consider what great benefits accrue not only to single persons, / but to whole communities, and sometimes even to nations by two or three vegetables, as many reptiles and insects, and as few minerals, of most of which the uses were unknown to the ancients, and those of the rest but little known ... The first of these is sugar, made of the express juice of the sugar cane, which is brought both from the *West* and *East Indies,* and that in such quantities, that the little island of *Barbados* alone furnishes *Europe* with the lading of many score ships in a year; and a sober knight which was governor of it, told me, that if I much misremember not the number, that small island did then furnish *England* with about eleven hundred thousand pound weight of sugar of its own growth. The other plant, though reckoned but a weed, makes at this day a great part of the commerce between *Europe* and some American country, especially *Virginia*; the latter of which country sends yearly to *England* alone a considerable fleet freighted almost only with tobacco ... All which improvements ought to excite man's gratitude to him, that made those creatures to his hand, and endowed him with a rational faculty and fit organs to exercise his plenary dominion over them.[59]

There could hardly be a richer description of the various dimensions of Boyle's vision of empire. Here England's colonies, both commercial and settlement, become the site for man to build his Adamic 'plenary dominion'. The abundant varieties of vegetables and minerals will aid man's restoration of his empire, and they will do so through providing knowledge about their characteristics, such as where they are grown and how they can be useful to man. Boyle's intentions here are clear: he aimed to legitimate English colonization by demonstrating that it was entirely consistent with the Adamic project of reaping the world's fruitfulness and re-establishing man's original empire over the earth. Any potential tension between the English and the universal character of the project was sidestepped by Boyle's underlying belief that England had assumed the Protestant mantle for mankind.

Having established the theological continuity between Boyle's natural philosophy and his interests in colonization, let us now examine Boyle's involvement in colonial institutions. The fact that Boyle was a shareholder in the Hudson's Bay Company is not widely written about. Yet in 1676 Boyle's name appears in the minutes of the Hudson's Bay Company as one of the thirty-four general stockholders of the company.[60] Six years later in 1682, his name appears on a similar list. Moreover, Boyle's membership enabled him to meet one of the Captains employed by Company, William Bond, whose observations Boyle recorded in his *Work-Diary XXXVI*.[61] The Governor of the Company and its settlements, John Nixon, was also an informant to Boyle and provided him with information about the variation of the compass needle in the area of Hudson's Bay.[62] The

Boyle Papers preserve a report from Nixon in which he details his wish to send men into the further reaches of the country.[63]

Perhaps one of the best contexts in which to examine the continuity between Boyle's theology and his colonial involvement is that of the attempts to proselytize indigenous peoples, whether they be Amerindians, the Irish or the Indians in India. Originally a Puritan endeavour during the Commonwealth, the Restoration endowed the Society for the Propagation of the Gospel, renamed the New England Company, with respectability. As James Jacob has suggested, this is perhaps one of the reasons Boyle was appointed as the Governor of the New England Company; he was a member of the established Church, as well as a close friend and ally of Clarendon.[64] In a letter to the Commissioners of the United Colonies of New England in May 1662, probably drafted by Henry Ashurst but approved by Boyle, he explained that the king had granted a charter for incorporating the New England Company. This would mean that 'many of the Nobility and other persons of Quality and most of those Gentlemen that were imployed in the like worke are Authorized and appointed to endeavour the carrying on that pious designe of converting the heathen natives'.[65]

In a letter to Michael Boyle, the Bishop of Cork from 1660 and then in 1662 the Archbishop of Dublin, Boyle reflected upon his appointment as Governor of the New England Company:

> His Majesty has been pleas'd without my seeking ... to appoint me Governour to a Corporation for the Propagating of the Gospell among the heathen Natives in New-England, and other parts of America. And this Corporation being at great charges for severall necessary Workes & especially for the translating & Printing of the Bible in the Indian tongue.[66]

Despite Boyle's seemingly virtuous ideals, we can glean a better understanding of the intentions of the New England Company from one of John Winthrop's letters to Boyle in 1662. In his letter, Winthrop put forward a number of proposals 'concerning the businese of imploying the Indians in our parts of New England'.[67] The aim was to treat the native Americans according to the same principles with which the English treated the land: improvement and cultivation:

> Those <westerne> parts of New England betweene the Narogansett Bay, & the River of New London are most populous of Indians, and a people more civill, & active, & industrious then any other of the adjacent parts: amongst them also there are a people, which live very neere the English, and doe wholly adheare to them, and are apt to fall into English imployment, therefore I have thought it an opportunity for the civilising of them & thereby the bringing them to hearken to the Gospell may be the easier effected.[68]

Winthrop described the 'benefitt to themselves' – that is, to the Indians – as:

1.   The civilising of them
2.   They would thereby be in a nearer way to <service> & the knowledge of the Christian religion, which is that great work this honorable corporation intends
3.   They would be furnished with such necessaries as may make their lives more comfortable, as civil people have.
4.   It would be a great benefit to the English people here, in a way of vending store of their commodities[,] especially drapery <of which now the Dutch have the greatest trade in those parts> for there be many thousands which would willingly weare English apparel if they knew how to purchas it, which must be easily done by the improvement of their owne labour in a due way; & besides many other manufactures would be vended.[69]

Here Winthrop's language is that of proselytizing, improvement and cultivation. Moreover, as Winthrop's fourth point makes clear, the purpose of the proposal to employ Amerindians was to exploit the New World – and in this context its people – in order to benefit England.

The evangelical dimension of Christianity, present long before the early modern era, manifested itself in the Elizabethan period during the struggle with Spain in the New World. In the mid seventeenth century the conversion of non-Christians became one of many Protestant projects for universal reformation. We recall here Cromwell's decision to allow the Jews back into England in 1656, an act influenced by the belief the conversion of non-Christians was a necessary pre-requisite to the second coming of Christ. As James Jacob has pointed out, 'one more aim Boyle shared with [Peter] Pett and the Hartlib Circle was that empire serve not only the joint purposes of science and trade but religion and reformation a well'.[70] The empire to which Jacob refers here is that of England's colonies. Jacob's claim for Boyle's 'imperialism'[71] is bold; that he held an 'aggressive, acquisitive, materialistic, imperialistic ideology justified in the name of Reformation'.[72]

I suggest that Boyle's interest in proselytizing had less to do with millenial ideas and more to do with a general religious conviction that teaching non-Christians about Protestantism would 'improve' their state. This sentiment had a particular resonance for Boyle as a Director of the East India Company because the reciprocal nature of trade, together with the reliance of English merchants upon the Indians, brought the needs of the Indians to light. In a letter to Major Robert Thompson, a member of the Court of Committees of the East India Company, as well as a member, and later Governor, of the New England Company in the 1690s, Boyle made it clear that he saw proselytizing as an integral part of maritime trade. He reminded Thompson that when he had sat on the Court of Committees of the East India Company, he

venturd to make a motion that some Course might be thought on of doing some considerable thing for the Propagation of the Gospell among the Natives in whose

Countreys we have flourishing Factories. And indeed it seemd to me very fit, that we whose endeavours God had of late so signally prosperd should pay him some visible acknowledgement of his many Blessings; and that remembring our selves to be Christians as well as Merchants, we should attempt to bring those Countreys some spiritual good things, whence we so frequently brought back Temporal ones.[73]

English colonies in which native peoples were converted to Protestantism, just as Boyle described in the letter above, were central to and consistent with his natural philosophy. Jacob's argument about Boyle's 'aggressive, acquisitive, materialistic, imperialistic ideology justified in the name of Reformation'[74] holds true to the extent that Boyle certainly brought together commercial, colonial and religious ideas. I suggest, however, that Boyle's conception of empire was more complex than his having simply an 'imperialistic ideology'. For Boyle, the term empire was still used only to denote man's prelapsarian dominion. It is more historically accurate to say that Boyle supported English 'colonies' (a term he *did* use), and that he legitimated this colonial project in terms of the idea of man's original empire over the world. The continuity between Boyle's theology and his interest in colonization was made possible by his adaptation of a language of universal Adamic empire to suit a very *English* colonial project.

Natural Philosophy and the New World

So far this chapter has explored the influence of Boyle's natural philosophy upon his interest in English colonization and trade. In this final section I investigate the reverse relationship. How did Boyle's colonial interests shape his natural philosophy? We have already witnessed Boyle's emphatic statements in the *Usefulness of Experimental Natural Philosophy*, among other works, that the New World would yield knowledge about a natural world unknown to the ancients and their Aristotelian intellectual descendants. Navigators and travellers had much to offer natural philosophers. In fact, Boyle argued, their testimony was vital to natural philosophers in the same way that the Christian Virtuoso depended upon the testimony of inspired men. From fairly early in his career, Boyle hosted such a variety of foreign visitors that John Aubrey observed, 'the reputation which he had acquired among foreign nations was so great, that no strangers, who came among us, and had any taste for learning and philosophy, left England without seeing him'.[75]

In the preface to *Observations and Experiments about the Saltiness of the Sea* (1673), Boyle told his audience that he had gleaned the information from 'Sea Captains, Pilots, Planters, and other Travellers to remote parts'.[76] Boyle's remarks are an excellent illustration of the way that his involvement in colonial institutions brought him distinct scientific advantages. His contact with these various travellers, Boyle wrote, was the fruit of his 'years a member of the Council appointed by the King of *Great Britain* to manage the business of all the *English Colonies* in the Isles and Continent of *America*, and of being for two or three years one of that Court of Committees (as they call it) that has the superintending of all the affairs of the justly famous *East-Indian Company of England*.[77]

Boyle's most explicit statement about the importance of testimony from travellers is to be found in Part 1 of the *Christian Virtuoso* where he draws a fascinating analogy between the reliability of knowledge derived from theological and natural philosophical informants. Boyle's argument is that the Christian Virtuoso must strive 'to Improve his Knowledge of Divine Things', and that one of the means of achieving this is to 'repose a great deal of Trust in the Testimony of Inspir'd Persons'.[78] The Christian Virtuoso should 'be allow'd to ground a Belief about such things, on the Relations and other Testimonies of those that were in the Scripture-Phrase, *Eye Witnesses and Ministers* of / the things they speak of', just as he who 'is curious to learn the State of that New-world' must consult 'with Navigators and Travellers to *America*'.[79] Information collected from the New World, then, was clearly a necessity to the natural philosopher. Boyle's description of the analogy is as follows:

> an ordinary Sea-man or Traveller, that had the opportunity with *Columbus* to sail along the several Coasts of it, [America] ... was able at his return to Inform Men of an hundred things, that they should never have learn'd by *Aristotle's* Philosophy, or *Ptolomy's* Geography ... And *as* one, that had a candid and knowing Friend intimate with *Columbus*, might better rely / on His Informations about many particulars of the Natural history of those Parts, than on those of an hundred School-Philosophers ... *so* ... may we rely on the Accounts given us of Theological Things, by the Apostles, and constant Attendants of him that lay in the *Bosom of God his Father*, and Commission'd them to declare to the World *the Whole Counsel of God*, as far as 'twas necessary for Man to know.[80]

In a description of the virtues of 'mixed mathematics', that is, what we might call 'applied mathematics', Boyle articulated his belief that travellers were vital to the natural philosophers' epistemic project of creating man's empire of knowledge:

> if Schollars and Travellers were more generally conversant, the History of Nature would be far better adorn'd with lively representations of Plants, Animals, Meteors, &c. and also *by* several parts of the Art of Navigation, and particularly that which they call *Histriodromia*, or the Doctrine of the Lines by which Pilots make their Ships to sail. Now if in these, and divers other Instances that may be given, it must be acknowledg'd, that mixt Mathematicks may be serviceable to the Naturalist, and assist him to promote the Empire of Man.[81]

Here, in the language of man's plenary empire, Boyle proposed a collaboration between scholars and travellers. Such a collaboration is very similar to the other major collaboration which Boyle proposed, that of naturalists and tradesmen.

Boyle's work on the History of Trades is well known. Originally a Baconian idea articulated in the *Advancement of Learning* and the *Parasceve*, the History of Trades was an attempt to make natural philosophy learn from manual arts.

Using agrarian metaphors, Boyle argued that a History of Trades was 'one of the best means to give Experimentall Learning both growth and fertility, and ... to prove to natural Philosophy what a rich Compost is to Trees, which it mightily helpes, both to grow faire and strong and to bear much fruit'.[82] Or, to put it another way, 'the Naturalist may probably advance Trades, and assist Man, by the blessing of the Author of Nature, to recover part of his lost Empire over the works of Nature'.[83]

Boyle systematically articulated the virtues of tradesmen, and why the naturalist could obtain from them 'informations, that may be very / usefull to his design'.[84] First, tradesmen are precise and careful; they 'are usually more diligent about the particular things they handle, than other Experimenters are wont to be'.[85] Second, because 'Necessity is the mother of Inventions' so they are 'very Industrious and inventive' and they therefore 'discover new uses and Applications of things'.[86] The third point about tradesmen is their ability to build new knowledge, 'unknown to Classical writers'.[87] Moreover, these inventions 'are not only factitious, but divers of them Natural'; that is, they produce knowledge about natural substances. Consequently, Boyle tells Pyrophilus, 'I learn'd more of the Kinds, Distinctions, Properties, and consequently of the Nature of Stones, by conversing with two or three Masons, and Stone-cutters, than ever I did from *Pliny* or *Aristotle*, and his Commentators.'[88]

Boyle's fourth point is that even though they are 'unacquainted with Books, & with the Theories & Opinions of the Schools', tradesmen are able to examine phenomena or objects with 'Mechanical waies' and this makes them 'the more new and Instructive, and consequently the more fit to be admitted into the History of Nature'.[89] Fifth, 'the Observations that Trades-men can supply us with, though they are not probably at any one time so accurately made by them, as they would be by a Learned man; yet that defect is recompensed by their being more frequently repeated, and more assiduously made, than most of the Experiments wherewith men of Letters have / furnished Natural History'.[90] Finally, Boyle argues that some tradesmen are able to observe new things which are 'unobserved by others, both relating to the Nature of the things they manage, and to the Operations performable upon them'.[91]

What is clear from Boyle's analysis is that underlying trades is an epistemology germane to natural philosophy. Because tradesmen rely upon practical experience and observation, their craft is experimental and creative rather than discursive and derivative. As Antonio Perez-Ramos has argued in his study of the tradition of 'maker's knowledge':

in Boyle's conception, the production of 'works' (he employs the Baconian word) places the natural philosopher in the position of Nature herself, as the craftsman who, by causing foreseeable effects to be produced, is supposedly learning the most secret workings of that arch-craftsman, *Natura faber*.[92]

Perez-Ramos's observation is especially pertinent to Boyle's work on the History of Trades because a trade is a practical activity through which man can reproduce (and thus 'know') God's works. So the epistemological novelty of natural philosophy – its experimentalism – extends to tradesmen, whose crafts Boyle sees as creative and inventive of practical knowledge.

It was not only tradesmen, but also traders – merchants – whose information Boyle conceived of as reliable. Unlike tradesmen, who were knowledge-makers, merchants did not share with natural philosophers the epistemological affinity with creative knowledge. Merchants traded rather than produced but, as Mary Poovey has shown, it was paradoxically because merchants were not practitioners of knowledge that Boyle conceived of them as reliable witnesses. Their credibility stemmed from their *disinterest* in knowledge.[93]

What implications did Boyle's epistemology have for information collected in the New World? A number of scholars, chiefly Steven Shapin and Simon Schaffer, have addressed the importance of the epistemology of Boyle's experimentalism. Steven Shapin has shown how Boyle wrote his experimental reports 'rich in circumstantial detail' which was intended 'to enable readers of the text to create a mental image of an experimental scene they did not directly witness'. This act Shapin terms 'virtual witnessing.[94] Boyle also avoided flowery prose. 'This plain, puritanical, unadorned (yet convoluted) style was identified as *functional* ... the "florid" style to be avoided was a hindrance to the clear provision of virtual witness: it was, Boyle said, like painting "the eye-glasses of a telescope".'[95]

Robert Iliffe, building upon the work of Shapin and Schaffer, has placed the 'codes proper to an English gentleman' in an international context and argued that Englishness was constructed in deliberate opposition to other forms of 'scholarly endeavour' characteristic of foreign nations.[96] With this international context in mind, I suggest that we can see how Boyle used the idea of England's colonies as a source of status for certifying the reliability of his witnesses.

An examination of Boyle's references to information gleaned abroad reveals that many of the informants had a connection with one of the English colonies or colonial companies; they were most often physicians, governors, members of colonial companies, preachers or merchants who travelled to the colonies frequently to trade. One of Boyle's most frequent sources of information was Henry Stubbe, Physician to the Governor of Jamaica between 1662 and 1664, but with whom Boyle later fell out.[97] When introducing testimony from Stubbe in either his published work or his work-diaries, Boyle nearly always made a point of making clear his colonial association with Jamaica. In a series of 'Strange Reports' appended to *Experimenta et Observationes Physicae*, for example, Boyle described Stubbe as 'An Inquisitive Gentleman lately Return'd from *Jamaica*, where he was Imploy'd by the Governour to make Discoveries of Natural Things'.[98] Stubbe told Boyle about the trees in Jamaica which bore 'Silken Cotton' which 'surpass in bigness and height the /larger sort of our English Oaks'.[99]

A good deal of the entries in Boyle's *Work-Diary XXI* are sourced from Stubbe.[100] Boyle entitled this particular work-diary his 'Outlandish Booke' and used it to record testimony from explorations into foreign countries. Information collected from Stubbe in these records of 'Promiscuous Experiments' includes that pertaining to scorpions, crocodiles and alligators, Jamaican mangrove trees, the nature of breezes in Barbados, the distance of salt water beneath the soil in Jamaica, and the skin colour of Africans.

Another of Boyle's most prolific informants was John Winthrop, Governor of Connecticut, whom Boyle described in the *Usefulness of Experimental Natural Philosophy*, Part 2, as 'an ancient *Virtuoso*, Governor to a considerable Colony in the Northern *America*'.[101] Boyle's asked Winthrop about 'the making of several kindes of Suger', and Winthrop replied by informing Boyle of various ways of making sugar, one of which he discovered through 'a very eminent and skilful Planter'[102] who tried to procure the liquid from corn. Boyle made a number of requests to Winthrop. In April 1664, for example, he asked to 'receive such an information of those severall particulars ... wherein the Naturall history of New england or any part of it differs from ours'.[103] The following year, he wrote once more to request Winthrop 'to send mee by parcels, as your leisure will permit, some Accounts of those particularitys, whether as to the Aire, or the Soyle, or the Husbandry, or the plants, or the Animals, or any other part of the Naturall History, (especially in relation to the Mineralls,) wherein your Colony, or any other part of New England that <you> are acquainted with, differs from other Countrys, especially those we here live in'.[104] In 1662, Winthrop attached to a letter a very lengthy description of various aspects of the natural environment of New England. Its subject matter included a physical description of corn (its 'Beautifull noble Eare ... Cloathed and Armed with strong thick huskes').[105] Boyle relied upon Winthrop's observations, for example, in his tract on 'Cosmical Suspicions', to confirm another report which he had received from New England about a change of climate.[106] Boyle recounts that he had 'the Honour to be standing by his Majesty when he received a solemne Addresse' from Winthrop, which corroborated the evidence about the increase in temperature from another of Boyle's correspondents.[107]

Boyle's work most heavily indebted to information gleaned in the colonies was his *General History of the Air*, which John Locke took care of publishing in 1692 after Boyle's death. In this text, we find numerous examples of Boyle making reference to information about various aspects of the climate in the English colonies in America. Boyle nearly always made a point of pointing out the colonial association of the informant. In the section on '*the* Heat and Coldness *of the Air*', for example, he states that 'Mr. *Nickson*, who was four Years Governour of the English Colony in *Hudson's Bay*, answered me, that when they sail'd within a certain Distance of floating Islands of Ice, if the Wind blew from thence toward the Ship, or as the Seamen, speak, if they were to the Leeward of the Ice, they could by the new and sensible Cold they felt, know that such Ice lay to Windward of them'.[108] Earlier, Boyle had referred to 'A learned Man that lived at *Jamaica*' who 'assured me, that when he laid in his Hammock, about three or four Foot from the Ground, though he had much Clothes under him, and little or none over him, he felt it cold beneath, and hot above'.[109] This led Boyle to conclude that 'the greater Heat that is usually found in our Air, during the Summer ... has manifest Effects upon such easily agitable Bodies as Liquors, and upon the Juices and Flesh of Animals'.[110]

One of the most fascinating examples of Boyle's use of colonial correspondents in the *General History of the Air* was his reference to two men from New Hampshire. We can trace this information right back to Boyle's notes from his interview with the men in his *Work-Diary*. In the *General History of the Air*, Boyle's reference to the two men is as follows:

> two Gentlemen belonging to the Province of *New-Hampshire* in New-England, (whence they came not long since) and imployed by that Colony to his Majesty here, answer'd me, that in the Winter the coldest Wind that blows in their Country, is the North-West; and being ask'd again, what was their hottest Wind in Summer, they told me, it was likewise the North-West.[111]

In Boyle's *Work-Diary* XXXVI, which is dated between the years 1685 and 1691, we can find his notes from the interview with these two men. It reads:

> the other day, *two Gentlemen* belonging to the Province of new Hampshire in New England (whence they came not long since) & imploy'd by that Colony to his Majesty here, answer'd me that in the Winter the Coldest Wind that blowes in their Country, is the Northwest: & being ask'd again, what was their hottest Wind in Summer, they told me it was likewise the Northwest.[112]

The fact that Boyle made good use of this particular interview illustrates the importance of men's colonial connections in establishing their reliability and trustworthiness. We note that Boyle chose to mention the fact that the two men were both 'gentlemen', and had been employed by His Majesty in the colony of New Hampshire.

From these examples it is clear that both in his published work and his *Work-Diaries*, one of Boyle's chief means of identifying his informants was their association with English colonies or colonial companies. Although it was necessary for Boyle to state the geographic origin of pieces of information, most of his descriptions of his informants extend beyond this. In *Work-Diary XXXVI*, for example, Boyle referred to Sir William Stapleton as 'a grave and experienc'd Gentleman who for many years has ... been Governour of several English Islands in America'.[113] The fact that Stapleton's information derived from America was a necessary piece of information, but his status as 'Governour of several English Islands' was not. Similarly, in *Work-Diary XXI*, Boyle reported having asked questions about the 'Silk-Cotton Tree' to 'The eminentest Person of one of our chiefe Plantations'.[114] Instead of naming him personally, or even naming the plantation with which he was involved, the information which Boyle believed established the reliability of this testimony was the status of his informant as the most eminent man in one of England's chief colonies.

Similarly, Boyle described Sir Edmund Andros, one of his informants in *Work-Diary XXXVI*, as 'An eminent person whose Curiosity & Affairs led him to visit divers parts of America (of one of which He was Governour)'.[115] Boyle's choice to include reference to Andros's governorship of the colony of New York (1674–81) was clearly to establish his status, since the information Andros provided actually pertained the 'Shining Flies' in the 'Islands of the Mexican Gulf'.[116] Later in the same *Work-Diary*, when Boyle mentioned his informant Colonel Richard Coney, he included no references to any gentlemanly qualities, presumably because the fact that he was '*The late Governour of the Bermudas*' was sufficient information to establish his reliability.[117] Moreover, instead of just stating the fact that Sir Hans Sloane was a naturalist in Jamaica, Boyle made a point of writing that he was 'imployd by *the Governour*' to make his 'discoveries of naturall things'.[118]

In some instances, Boyle would not name the informant but instead state his imperial association. In his third aphorism in the second part of the *Christian Virtuoso*, for example, Boyle discussed a number of creatures who were 'subjected to man's dominion, that are of very great use to him without exceeding the number of twenty or thirty'. He refers to the discovery of the mandioca 'whose root ... is believed to be poisonous; and a general of the English forces in *America* told me, that in his army several animals were killed by the use of it'.[119] Similarly, in the *General History of the Air*, Boyle mentioned having asked 'an intelligent Person that liv'd a good while in *Guinea*, how they did to keep their Water cool in so hot a Place'.[120] Later in the work, Boyle referred to the 'intelligent Person that was for many Years Consul of the English Nation at *Tripoli* in *Barbary*; and at / another time Governour of the *Castle* (called) *of the Coast*, belonging to the English African Company in *Guinea*'.[121] Relying so much upon testimony from employees of colonial companies, and also upon the merchants and Governors of colonies, Boyle was able to use England's colonial and commercial endeavours as a means of establishing the reliability of his informants.

I began this chapter by asking whether there might be a connection between Boyle's colonial interests and his natural philosophy. The answer to that historical question is illuminating. Boyle conceptualized English colonies as cultivating the earth's resources, and thus they were part of the project to re-establish man's empire over the creation.

Boyle's work marked a significant moment in the intellectual origins of the British Empire. In his detailed theological defence of experimental philosophy, he emphasized the importance of the agrarian dimension of Adam's original empire; the command over the fruitfulness of the world. In doing so, Boyle laid the groundwork for a new justification for colonization: England could legitimately cultivate the agrarian fruitfulness of any empty soil because this was God's original imperial command to mankind. It was John Locke who ultimately made this justification for possessing colonial property. By appropriating the universal language of mankind's original empire to suit a very English colonial project, Boyle prefigured Locke's theory of property in the *Two Treatises of Government*.

4 THE ROYAL SOCIETY AND THE ATLANTIC WORLD

So good an opportunity as this I could not let passe without putting you in mind of yr being a Member of ye Royall Society, though you are in New-England; and even at so great a distance, you may doe that Illustrious Company great Service ... [by] communicating to them all the Observables of both Nature an Art, yt occur in the place, you are ... Sr, you will please to remember, that we [the Royal Society] have taken to taske the whole Universe ... It will therefore be requisite, that we purchase and entertain a commerce in all parts of ye world.

Henry Oldenburg to John Winthrop, Governor of Connecticut, 13 October 1667[1]

In October 1667, Henry Oldenburg, the Secretary of the Royal Society of London, wrote to John Winthrop Jr, the Governor of Connecticut, reminding him of his responsibility to help the Society *lay open ... an* Empire *of Learning* as Edmond Halley put it in the preface to the *Philosophical Transactions* in 1686.[2] In the forty years following the Restoration, Winthrop was one of many correspondents of the Royal Society, sometimes Fellows themselves, who would send back 'rarities', 'curiosities' and detailed knowledge from the colonial periphery to London. This transfer of knowledge was tangible and haphazard; letters and wooden boxes were shipped across the Atlantic. The former recorded natural histories of places throughout the Americas and observations of weather patterns, while the wooden boxes contained berries, soil samples, rocks and occasionally even animal specimens.

This chapter explores the colonial dimension of the Royal Society's extensive correspondence with men throughout the New World. The colonial context is particularly illuminating: the Society conceived of correspondence with the New World as a vital part of its project to restore man's epistemic dominion over nature.

The Royal Society engaged in two practices of knowledge collection and organization. The first was the attempt to create an encyclopedic natural history; a practice which relied heavily upon information sent from England's Atlantic and Caribbean possessions. The second practice was the creation of the Society's repository of specimens. The repository, I will argue, aspired to be a tangible microcosm of an empire of knowledge. As Robert Hooke put it, it was 'as full and compleat a Collection of all varieties of Natural Bodies as could be obtain'd'.[3] Although the success of the repository was limited, it remains an important feature of the early Royal Society's endeavours because it was the focal point of a new vocabulary of natural philosophical collecting.

A Universal Natural History

Compiling an encyclopedic natural history was an ambitious project. Nevertheless, it was a popular one; many learned men with an interest in natural philosophy were eager to glean as much information as possible about the New World. In a letter to the mathematician René François Sluse in 1667, Oldenburg explained the daunting enormity of the Baconian project:

> For we seek thoroughly to scrutinize everything – the heavens, the Earth, the subterranean world; the air, the meteors, and stars; rivers, seas, vegetables, minerals, animals; so that there is nothing of all this that you may not explore, wherever you turn.[4]

Oldenburg's ambition was made possible by his confidence that like-minded men – 'lovers of truth' – would offer their services to the Royal Society as correspondents. He advocated:

> the careful compilation of a Natural History such as our illustrious Bacon formerly proposed ... We urge everyone who is talented, skilful, partial to philosophy, and a lover of truth, wherever in the world he lives, to join hands with us and furnish us with accurate reports of all that is worthy of note occurring in the region where he lives.[5]

England's New World colonies were of course not the only places with which the Society communicated. An examination of the Oldenburg correspondence, for example, illustrates the extent of Oldenburg's own, and the Society's, European correspondence. The New World, however, did play a significant role in the Society's information-gathering projects. This was in no small part due to the fact that a number of English colonists were eager to be correspondents. John Winthrop was a Fellow of the Royal Society, and his fellow colonists in New England demonstrated an abiding interest in natural philosophy. The Congregationalist Minister in Massachusetts, Increase Mather, established a similar society, the Boston Philosophical Society, in 1683, and a number of men such as the travel-writer John Josselyn and the natural philosopher Richard Ligons wrote natural histories of New England and Barbados respectively.[6]

Founded in 1660 and given its second Royal Charter in 1663, the Royal Society was in some ways the fulfilment of the Baconian dream: a public, scientific institution dedicating itself to the advancement of natural knowledge for the good of mankind. As Michael Hunter has pointed out, it 'represented a new type of institution, a public body devoted to the corporate pursuit of scientific research, something unprecedented whether in this country or elsewhere'.[7] The Society was given the right to license its own books, possessed its own coat of arms and enjoyed the patronage of Charles II. Yet, despite these ostensible symbols of success and establishment, the Society was forever short of funds. Charles II acted as patron, but he endowed the Society with no money. Instead, members were charged admission fees and weekly dues which many neglected to pay.

Although the Society proudly proclaimed a unifying motto, *Nullius in Verba* (roughly, 'on no man's word'), it would be wrong to assume that all members of the Royal Society shared the same ideas about the Society's projects. The membership of the early Royal Society was varied, socially as well as on religious grounds.[8] One of the two original secretaries, Henry Oldenburg, bore more resemblance to an 'intelligencer' in the Hartlibian tradition – with his extensive European correspondence and connections – than he did to a Christian Virtuoso in the mould of Boyle, for example.[9] Oldenburg's personality and ambition towers over the early history of the Royal Society. The *Philosophical Transactions*, for example, were his pet project, and he maintained a prolific correspondence with men throughout Europe and the New World. It was quite possibly his correspondence that fuelled accusations of his involvement in 'dangerous designs and practices' and saw him spend two months of 1667 imprisoned in the Tower of London.[10]

The Society spent the Restoration engaged in an ongoing project to legitimate both its own existence as well as the virtues of natural philosophy. Thomas Sprat's *History of the Royal Society of London* (1667) was instrumental in this project. Sprat constructed a foundation narrative about the Royal Society that downplayed its members' Puritan connections and emphasized instead the formative role of royalist Fellows such as the poet Abraham Cowley. This was clearly a political move, designed to marginalize alternative stories about the Society's origins, for example Comenius's idea that the Society inherited the aims of the Bohemian tradition of learned academies.[11] It was not the radical Protestantism of the Invisible College which Sprat established as the Society's precursor, but the peaceful – and royalist – spires of Oxford in which the Philosophical Club met. Such palatable political and religious associations helped bolster Sprat's defence of the Society against accusations of irreligion.

Sprat's ideological project was to construct experimental philosophy as a means of producing knowledge that could command assent; an epistemological prevention of the kind of bloody and violent disagreements which paralysed England during the Civil Wars. Put another way, the Royal Society was the institutionalization of an attempt 'to create a domain of knowledge production *outside* of political and theological discussion'.[12]

Producing disinterested knowledge was one of the Royal Society's most important aims, and it is in this context which we should view the project to write a universal natural history. As Mary Poovey has shown, universal knowledge about the world was made possible once men could put behind them the 'interested' and 'contingent knowledge associated with reason of state or the instrumental knowledge associated with an individual's monetary gain'.[13] As we saw in the case of William Petty, however, *claims* to disinterested knowledge were frequently rhetorical, and concealed the interests, financial or otherwise, of the person in question. Petty, we recall, had a great financial stake in Ireland, despite the objectivity claims of his political arithmetic.

We can make an analogous point about the Royal Society's concern with compiling a universal natural history. Despite the Society's contention that such universal knowledge was free of theological and political debate, the natural history project itself was motivated by the theological belief that man should reconstruct Adam's plenary empire. Indeed, this religious belief was one of Sprat's main weapons against accusations of heretical ideas. Like Boyle, Sprat was adamant that in fact natural philosophy brought man closer to God by enabling him to understand and appreciate God's power as it worked through the Creation. 'There can be no just reason', he argued, 'why an *Experimenter* should be prone to deny the essence, and properties of *God*, the universal Sovereignty of his *Dominion*, and his *Providence* over the *Creation*.'[14] Far from elevating himself to a position in which man does not need God, using natural philosophy to understand the complexity of the natural world will humble man and enable him to appreciate God's power:

> he will best understand the infinit distance between *himself*, and his *Creator*, when he finds that all things were produc'd by him ... This will teach him to *Worship* that *Wisdom*, by which all things are so easily sustain'd, when he has look'd more familiarly into them, and beheld the chances, and alterations, to which they are expos'd. Hence he will be led to admire the wonderful contrivance of the *Creation*; and so to apply, and direct his praises aright: which no doubt, when they are offer'd up to *Heven*.[15]

Indeed, appreciating God in this way, through the study of nature, was 'the first service, that *Adam* perform'd to his *Creator*, when he obey'd him in mustring, and naming, and looking into the *Nature* of all the *Creatures*'.[16] Sprat's conclusion from this argument is a Baconian one that legitimates the systematic study of nature: 'So true is that saying of my Lord *Bacon*, *That by a little knowledge of Nature men become Atheists; but a great deal returns them back again to a sound Religious mind*.'[17]

Conceiving of natural knowledge in terms of man's original and normative relationship to the earth was one of the most important ways of justifying the new philosophy. The antiquarian Elias Ashmole is one example of a Royal

Society Fellow who articulated his conception of natural knowledge in Adamic terms. In his alchemical treatise, Ashmole stated that Adam

> was so absolute a *Philosopher*, that he fully understood the true and pure knowledge of *Nature* (which is no other then what we call *Naturall Magick*) in the highest degree of Perfection, insomuch, that by the light thereof, upon the present view of the *Creatures* he perfectly knew their *Natures*, and was as able to bestow names suitable to their *Qualities* and *Properties*.[18]

Ashmole went on to explain:

> the *Magick* here intended, and which I strive to Vindicate, is *Divine, True*, of the *Wisdom* of *Nature*, & indeed comprehendeth the whole *Philosophy* of *Nature*, being ... a *Perfect Knowledge of the works of God, and their Effects*. It is that, which ... *reduces all* naturall Philosophy *from variety* of Speculations *to the* magnitude of workes, and *whose* Misteries *are far greater then the* naturall Phylosophy *now in use and reputation will reach unto.* For by the bare application of *Actives* to *Passives* it is able to exercise a kind of *Empire* over *Nature,* and work *wonders.*[19]

One aspect of the ideal of recreating man's empire over the natural world was the sense of the inadequacy of the existing state of knowledge. Joseph Glanvill, the Anglican clergyman and staunch supporter of the Royal Society, argued that the Society would produce a history of nature superior to those preceding it:

> The *Histories* of *Nature* we have *hithero* had, have been but an *heap* and *amassment* of *Truth* and *Falshood, Vulgar Tales,* and *Romantick* Accounts ... But now, the frame of this *Society* suggests excellent ground to hope from them *sincere* and *universal Relations,* and the best *grounded* and most useful Collection of the Affairs of *Art* and *Nature,* that ever yet was extant: And as they have *peculiar* Privileges for the gathering the *Materials* of *Knowledge,* so They have the *same* for the *impartment* and *diffusion* of them.[20]

Glanvill was specific about how the Royal Society should achieve this universal natural history. He wanted:

> to make a *Bank* of all the *Vseful Knowledge* that is among Men: For either by their *whole Body,* or *some* or *other* of their *particular* Members, they hold a *Learned Correspondence* with the greatest *Virtuosi* of all the known *Vniverse,* and have several of their own *Fellows* abroad in *Forreign* Parts, by reason of whose *Communications,* they know most of the *valuable Rarities* and *Phaenomena* observed by the curious in Nature.[21]

What distinguished the Royal Society's project from that of its predecessors was not only its universalism, but also the systematic and organized way in which it was pursued. As Oldenburg put it, 'a truly natural history should be put together, one that is exact and compiled with the greatest care and accuracy in order that in the end a solid and useful philosophy may be based upon it'.[22] Systematization was necessary in order to render the knowledge useful: 'All lovers of true science will unanimously and gladly agree to its adopting their respective labors to itself for a profitable purpose.'[23]

The most novel feature of the Society's pursuit of natural history, therefore, was methodology. In contrast to Aristotelian textual exegesis, experimental philosophy was open and public in its attainment of truth. This public creation of reliable knowledge was the Royal Society's *raison d'être*, because, as Oldenburg explained in his best Baconian voice, 'the improvement of all usefull Sciences and Arts' was not to be through 'meer speculations' but rather through 'exact and faithfull Observations and Experiments'.[24] It is unsurprising, then, that the project of compiling a universal natural history was to be carried out systematically, often through a series of instructions from the Royal Society on methodology.[25]

Sprat praised this characteristic of the Royal Society's natural philosophical investigations highly. Describing the way in which the Society pursued its quest for knowledge, Sprat writes:

> First they require some of their particular Fellows, to examine all Treatises, and Descriptions, of the Natural, and Artificial productions of those Countries, in which they would be inform'd. At the same time, they employ others to discourse with the Seamen, Travellers, Tradesmen, and Merchants, who are likely to give them the best light.[26]

Sprat was quite specific when it came to describing the Society's method of enquiry:

> They have compos'd Queries, and Directions, what things are needful to be observ'd, in order to the making of a Natural History in general: what are to be taken notice of towards a perfect History of the Air, and Atmosphere, and Weather: what is to be observ'd in the production, growth, advancing, or transforming of Vegetables: what particulars are requisite, for collecting a compleat History of the Agriculture, which is us'd in several parts of this Nation.[27]

A close reading of the *Philosophical Transactions* reveals that in almost every edition of the journal between its first publication in 1665 and 1700, there was at least one report pertaining to the natural environment of the New World. The format and style of these reports varied. The most common were copies of letters sent to the Society from correspondents in the Americas. John Clayton, the Rector of Croston at Wakefield, for example, sent a letter 'giving a farther account of the Soil, and other Observables of Virginia'.[28] In 1695, Sir William Beeston, the Governor of Jamaica, had a letter published which contained his observations about the barometer and of a hot bath in Jamaica.[29] John Winthrop wrote to the Royal Society many times describing natural artefacts of New England in detail. Speaking of his recently acquired knowledge, Winthrop reported: 'I know now, whether I may recommend some of the productions of this Wilderness as rarities or novelties ... There are ... small Oaks, which though so slender and low ... have yet Acorns and cups upon them.'[30]

A second type of report regarding the New World was the letters from Fellows within England recounting information they had been sent from abroad. When John Locke, in his capacity as Secretary to the Board of Trade, came across information about the natural history of the Americas he copied it out and forwarded it to Oldenburg, who as the editor of the *Philosophical Transactions* increasingly became a conduit for such information. In May 1673, for example, Locke sent Oldenburg 'an account I lately received from New: Providence one of the Bahama Islands concerning fish there'.[31]

Another Fellow of the Royal Society, the physician and naturalist Martin Lister, also sent information he received to Oldenburg. Lister's correspondent was one Mr Thomas Townes, who was born in Barbados and returned there after matriculating at Christ's College, Cambridge, in 1664. In 1673, Lister wrote to Oldenburg and transcribed 'an observation or two' which Townes had sent him:

> I have heard it questioned, whether America have not some plants common wth those of Europe, especially ye more Northern parts of it. The soil here is fertile, though not above a foot or two thick upon a whole & spongie lime- stone rock, whch affords good Quarries here & there ... Indeed ye whole Island appears in a manner like a scattered town, wch wth ye perpetual green fields & woods makes this place very pleasant.[32]

A third type of entry in the *Transactions* which dealt with America was the book review. In 1672, for example, there is a review of a book entitled, *The American Physitian; or a Treatise of Roots, Plants, Trees, Shrubs, Fruit, Herbs &c growing in the English Plantations in America*. The reviewer writes that the book is 'of good use, forasmuch as it may make a part of the Universal History of Nature'.[33] In most editions of the *Philosophical Transactions* similar reviews can be found, often dealing with botanical discoveries in the Americas.

The New World's botany was of particular interest to a number of Fellows, one of whom was the physician, naturalist and later the founder of the British Museum, Sir Hans Sloane. Like his colleagues, Sloane was motivated by a sense of the inadequacy of existing natural histories. In 1687, Christopher Monck, Duke of Albemarle and Governor of Jamaica, offered Sloane the position of his personal physician. Sloane set sail to the New World via Madeira and the Canary Islands. The party arrived in Jamaica in December that year, and they remained there until October 1688 when the Duke died. Sloane departed for England in March the following year, carrying with him approximately 800 plant specimens. The other fruit of Sloane's journey was his natural history entitled *Catalogus Plantarum quae in Insula Jamaica Sponte Proveniunt*. It was dedicated to both the Royal Society and the Royal College of Physicians and was published in 1696. Sloane also published a more general work of natural history about his voyage, entitled *Voyage to the Islands of Madera, Barbados, Nieves, S Christophers and Jamaica, with the natural history of the last of those islands; to which is prefixed an introduction, wherein is an account of the inhabitants, air, waters, diseases, trade &c.*[34] Sloane's collections, which by his death numbered above 71,000 objects, a library and herbarium, eventually became the foundation for the British Museum, first opened to the public in 1759.

Although most of the Society's Fellows never travelled to the New World, correspondence could be extremely useful. Oldenburg, for example, never travelled beyond Europe, but he was assiduous in his requests to John Winthrop. In 1671, for example, Oldenburg implored Winthrop to compose:

> a good History of New England, from the beginning of ye English arrival there, to this very time, containing ye Geography, Natural Productions, and Civill Administration thereof, together wth the Notable progresse of the Plantation, and the remarkable occurrences in the same.[35]

Similarly, writing to Richard Norwood in Bermuda, Oldenburg asked him 'to send in what Observables you might meet'.[36] Richard Stafford, who arrived in Bermuda in 1626, and was briefly the sheriff there, wrote to inform Oldenburg that he should receive from Captain Thomas Morlie 'the commandr: of our Magazieene ship such thing as I could at present procure'.[37]

In Oldenburg's requests for information we see some of the best illustrations of the Society's methodological emphasis upon the systematization and categorization of information. In the eighth issue of the *Transactions*, published in 1666, the astronomer Laurence Rooke devised a set of 'Directions for Sea-men, bound for far Voyages'.[38] They included, in part:

> 3. To remark carefully the Ebbings and Flowings of the Sea, in as many places as they can, together with all the Accidents, Ordinary and Extraordinary of the Tides; as their precise time of Ebbing and Flowing in Rivers, at *Promontories* or *Capes*; which way their current runs, what Perpendicular distance there is between the highest Tide and lowest Ebb, during the Spring Tides and Neap-Tides; what day of the *Moons* age, and at what times of the year, the highest and lowest Tides fall out: And all other considerable Accidents, they can be observe in the Tides, chiefly near Ports, and about Ilands, as in St. *Helena's* Iland, and the three rivers there, at the *Bermudas* &c.[39]

Five months after these instructions were printed, Robert Boyle set out the '*General Heads for a* Natural History of a Countrey, *Great or Small*'[40] in the *Philosophical Transactions*:

> *First,* in the Earth *it self,* may be observ'd, its dimensions, situation, East, West, North, and South: its Figure, its Plains, and Valleys, and their Extent; its Hills and Mountains, and the height of the tallest, both in reference to the neighbouring Valleys or Plains, and in reference to the Level of the Sea ... Whether the Countrey be coherent, or much broken into Ilands. what the nature of the Soyle is, whether Clays, Sandy ...

As also, by what particular Arts and industries the Inhabitants improve the Advantages, and remedy the Inconveniences of the Soyl ... *Secondly*, above the ignobler *Productions* of the Earth, there must be a careful account given of the *Inhabitants* themselves, both *Natives* and *Strangers,* that have been long settled there: And in particular, their Stature, Shape, Colour, Features, Strength, Agility, Beauty, (or the want of it) Complexions, Hair, Dyet, Inclinations, and Customs that seem not due to Education.[41]

Boyle's detailed instructions continued for several pages. The next volume of the *Transactions* published inquiries relating specifically to different countries. '*Inquiries For* Virginia *and the* Bermudas', for example, read:

1.   Concerning the Varieties of Earths;'tis said, there is one kind of a *Gummy* consistence, white and cleer: Another, white, and so light, that it swims upon water: Another, red, call'd Wapergh, like *Terra Sigillata*. Quaere, what other considerable kinds are there? And to send over a parcel of each.
2.   What considerable Minerals, Stones, Bitumens, Tinctures, Drugs?
3.   What hot Baths, and of what Medicinal use?
4.   What is the Original of those large Navigable Rivers, which empty themselves into the Bay of *Chesapeak*? And whether on the other side of that ridge of Mountains, from which they are supposed to proceed, there be not other Rivers, that flow into the *South-Sea*?[42]

The next series of inquiries was directed at 'Guiana and Brasil' and included questions about 'whether upon the Leaves of that *Brasilian* Tree, call'd *Cereiba*, there is, in a Sun-shiny day, found a *White Salt*, in that quantity, that one may gather as much from two or three Leaves, as will well salt a good pot of Broth?' There were so many inquiries, in fact, that '*The other Inquiries, ready for the other Countries above-named, are, to avoid tediousness, referred to [at] another opportunity*'.[43]

The following year, the introduction to the *Philosophical Transactions* proudly proclaimed that 'even our *former Tracts* ... have already brought in several pertinent *Answers; viz* from a *Sea-Voyage*, the *Caribbe*-Islands, and *Jamaica*'.[44] In that edition of the journal, there followed '*Enquiries and Directions For the* Ant-*Iles*, or Caribbe-*Islands*'.[45] The section 'Of Vegetables', for example, included the question 'whether the Juice of the Fruit of the Tree *Junipa*, being as clear as any Rock-water yields a brown Violet-dye?'[46] The queries are too numerous to quote at length, suffice to say that for the Antilles and Caribbean, they continue for six pages.[47]

Although the *Philosophical Transactions* portrays the aims and processes of the Royal Society's imperial epistemic project as straightforward and even well organized, the reality was rather different and more complex. The success of the Society's inquiries about natural knowledge was open to debate. During the first three decades of its existence, the Royal Society received a fair amount of criticism, and even mocking, for its iconoclastic adventures in the reformation of natural philosophy.

One of the most vehement critics, the physician Henry Stubbe, believed that the Society knew far less about the New World than it purported to:

> Just so when I went to Iamaica and desired that Honourable Personage Mr. Robert Boyle to procure some directions for Philosophical Inquiries in that Countrey; He, with blushing and disorder, tendered me from them a ridiculous paper which concerned most some particularities of China, and those Oriental parts.[48]

The Society's apparent lack of knowledge of geography was the subject of Stubbe's jousting in *The Lord Bacons Relation to the Sweating-Sickness Examined*, in which he argued that the Royal Society's project of natural history was 'not necessary to the World' unless one has 'occasion to send to the East-Indies to know what grows in America, or to Southwales for an account of Nova Zembla, or the Countries subject to the North and South Pole'.[49] More biting was Stubbe's question as to 'who can with any patience read how this famous Society sent to the Governor of Batavia in the East-Indies to know what grown in Mexico in the West-Indies?'[50] Although Stubbe's personal animosity towards the Royal Society colours his writing, his sentiments illuminate the fact that, lurking beneath the frequently triumphal tone of the *Transactions*, boastful of new discoveries to aid the compilation of a universal natural history, a number of problems plagued the process of accumulating knowledge. These issues are set in relief by the second practice of the Royal Society's epistemic empire-building: the collection of rarities.

The Royal Society's Repository: A New Vocabulary of Museums

After almost a decade of sending entreaties to Connecticut, in the autumn of 1669 Oldenburg finally received what he had been hoping for. Several wooden boxes arrived from New England containing a myriad of natural specimens. The sender was John Winthrop, who described the objects in an accompanying letter:

> There is in a broad round box a strang kind of fish wch was taken by a fisherman [in] ... Massachuset Bay in New England ... There is in an other box a fish wch is full of prickles wch they call a seahedghog; as also a small flying fish ... these flying fishes are chased by the dolphin, & that causeth them to fly out of the water ... in those seas betweene these parts and the West-Indies ... In the same box are heads of a vegetable we call silke grass ... pieces of the barke of a tree wch growes at Nova Scotia ... the eares of Indian corne ... There is also put aboard [the ship] loose ... the head of a deare wch seemeth not an ordinary head. It was brought far out of the country by some Indians.[51]

The second practice with which the Royal Society attempted to recover man's epistemic empire in America was tangible. It was the accumulation and display of specimens. In its drive for an encyclopedic representation of the natural world, a repository or museum was a microcosm of the ideal of an empire of knowledge. In the words of Robert Hooke, it was to be 'as full and compleat a Collection of all varieties of Natural Bodies as could be obtain'd', and, as Sprat optimistically stated, the Society had 'drawn together into one Room, the greatest part of all the several kinds of things, that are scatter'd throughout the *Universe*'.[52]

The practice of collecting natural specimens was not new to the seventeenth century. Medieval interest in the natural world derived from the belief that nature held allegorical keys to understanding God's two books: his word (the Bible) and his work (the Creation). One of the defining characteristics of medieval and Renaissance natural histories was their concern with the heuristic significances of animals and plants. Herbs as well as animals were of particular interest in the Renaissance, which is evidenced by the genre of herb collections called the *res herbaria*. Interest in the variety of herbs did not necessarily mean that an attempt was made to cite the exact location of a particular herb, or its variety. In fact, scholastic writers resorted to ancient categories for describing herbs, even when the herbs were sourced in the New World.[53]

This reliance upon ancient authorities for the interpretation of the natural world was accompanied by a view of the significance of nature as emblematic and laden with symbols of the divine. Collections of natural and artificial objects, just like natural histories, were seen as a representation of God's power to intervene in natural processes and produce miracles.[54] In Foucauldian terms, this is often seen as a distinctly pre-modern conception of nature because its heuristic principle was that of symbolic reference. In an epistemic shift, it was eclipsed by the taxonomic description characteristic of modern science.[55]

This epistemic transformation is perhaps easy to overstate, but there were some important associated changes in the practices of natural history writing and collecting in early seventeenth-century England. The first was an increasing interest in collecting foreign natural specimens, particularly those of the New World. A good illustration of the increasingly important role travel played in collecting is the case of the John Tradescants, senior and junior. Tradescant Sr was the gardener to Charles I and amassed a great collection of plants. Although he possessed American specimens, he had not deemed it necessary to travel to the New World himself. By contrast, several decades later his son John Tradescant the younger (*c.* 1608–62), whose collections formed the basis of the Ashmolean Museum, realized the importance of the New World to natural knowledge. He visited Virginia three times: first in 1637, then in 1642 and finally during 1653–4. The Calendar of State Papers records: 'In 1637 John Tradescant was in the [Virginia] colony, to gather all rarities of flowers, plants, shells &c.'[56] America was fast being imagined as a new Eden; a storehouse of the unknown natural world ready for rediscovery and classification.

In 1656, Tradescant Jr published a catalogue of his collection entitled *Musaeum Tradescantianum*. Funded by Elias Ashmole, it was the first catalogue of a museum printed in England, but its categories and classifications were only a semblance of order. The culture of collecting in mid seventeenth-century England remained largely in the form of cabinets of curiosities, categorized not by type or geographic location but often by the material from which they were made. The purpose of most collections was not to improve knowledge of the natural world but to represent social status through the conspicuous display of rare objects.

As we saw in Chapter 2, Francis Bacon's reform of natural history was founded upon the idea that its purpose was not to indulge curiosity but to advance knowledge. Bacon's proposed reforms were intimately connected to the disdain he felt for the unsystematic and amateur way in which collecting was practised in the Elizabethan and Jacobean courts, which he saw as one of the weaknesses of the existing state of knowledge. Collecting specimens of natural history was a popular pastime of the aristocracy, but Bacon viewed such collections as focusing only upon the curiosity of the artefact; they were aimless, the objects were decontextualized, and the trivial collections did not advance knowledge of their subject. Meticulous and empirical natural histories, Bacon argued, would be one of the most fruitful ways of improving knowledge. The Royal Society aspired to Bacon's ideals and hoped that its repository would be a tangible microcosm of man's plenary epistemic empire.

The Society's collection of curiosities existed from its earliest days, in fact, before the Society was formally incorporated. In 1662 it established a repository at Gresham College, London, which was attended by Robert Hooke who became known by the title 'Curator of Experiments'. In July 1678 the botanist Nehemiah Grew was asked to catalogue the repository. A year later the result was printed, and titled *Musaeum Regalis Societatis, or, A Catalogue and Description of the Natural and Artificial Rarities Belonging to the Royal Society and Preserved at Gresham Colledge*.[57] Grew's choice of the word 'museum' to describe the repository is significant. The *Oxford English Dictionary* reveals that the word began to be used in English only in the 1650s. By naming the Royal Society's repository a museum, Grew helped sanction the term for scientific enterprise.

In his catalogue, Grew attempted to realize the Adamic ideal of knowledge in which plants had names that corresponded to their natures. He tells us that, where plants and animals did not have names, he named them himself. In doing so, he tried to name them based upon 'some-thing more observably declarative of their Form, or Nature. The doing of which, would much facilitate and Improve the Knowledge of them. For so, every Name were a short Definition.'[58] In the last few paragraphs of his Preface, Grew places the cataloguing project in a Baconian context in which natural specimens contribute to the usefulness and improvement of knowledge: 'The greatest Rarity, if once experienced to be of good use, will soon become common.'[59]

The American colonies played a central role in the project of collecting speci-mens. Many of the Society's most reliable correspondents were stationed in the colonies and, furthermore, the colonies' position on trade routes facilitated the trans-Atlantic shipping of artefacts. Approximately two-thirds of the objects in Hans Sloane's personal collection, for example, had an American provenance,[60] and it was Sloane's colonial activity as physician to the Governor of Jamaica that enabled him to gather the eight hundred plants and menagerie of animals which became the basis for his collection.

Despite the relentless requests for New World knowledge, the Society's repository did not live up to Oldenburg's Baconian hopes. There were a number of reasons for this, not least the insufficient resources for the repository, but also the differing expectations about whether specimens collected should be held in common by the museum, or privately by individual members. The Baconian ideal, of course, entailed that specimens be held and used by the Society as a whole, but a number of Fellows owned their own private collections.

In fact, the collaborative and professional aspirations of the Society's reposi-tory were the exception rather than the rule. Elias Ashmole's story is a case in point. The son of a Staffordshire saddler, Ashmole rose through the social hierar-chy by trying to ingratiate himself with the Order of the Garter. He also married wisely. His second wife, Lady Manwaring, was considerably wealthy. She was also twenty years his senior. The fact that one of her sons tried to murder him may well indicate that his motives were showing. His shady marriage aside, the chief way in which Ashmole tried to promote himself was through his collecting. His collection, which we know today in the form of the Ashmolean Museum, originally belonged to the Tradescants, and it contained many American rarities. Arthur MacGregor's research into the surviving collections at the Museum has yielded a list of extant artefacts that can be traced to the 1685 catalogue. Items from the Americas include a paddle, ball-headed clubs, non-ball-headed clubs, a self-bow and a beaded wampum belt ('wampum' being a colonial abbreviation of the Algonquian word 'wampumpeake' which refers to white and purple tubular shell beads). There was also a shirt made of skin (the mantle of Powhatan, the Amerindian 'king' of Virginia which is decorated with beadwork), a skin pouch, a jaguar-tooth pendant, a string of ocelot teeth (ocelot is a native wildcat) and a hammock. In addition, a number of ethnological specimens included wampum beads, a model canoe probably from Canada, a South American Indian club, a hammock from British Guiana made of cords and twisted grass, and a hammock from South America made from white cotton thread.[61]

From about 1650 onwards, we know that Ashmole and his first wife ingrati-ated themselves with the Tradescants. Ashmole subsequently played a role in the publication of Tradescant's museum catalogue *Musaeum Tradescantianum*, and he was later – at least so he claimed – given Tradescant's collection as a gift.

Mrs Tradescant, perhaps unsurprisingly, disagreed about this.[62] The Tradescants had put effort into building the collection, but it was Ashmole who created from Tradescant's Ark an institution of considerable status. A new building was constructed for the museum in Oxford, and James Duke of York officially opened the *Museum Ashmoleanum* in 1683. Entry to the museum was open to the public.

The fact that Asmole's institution was called a museum is significant. Tradescant's *Musaeum Tradescantianum*, the catalogue of specimens which Ashmole assumed, was the first use of the term 'museum' as 'a building or institution in which objects of historical, scientific, artistic, or cultural interest are preserved and exhibited'.[63] The modern idea of the 'museum' entered the English language during this historical moment. That it featured in the title of Grew's *Musaeum regalis societatis* (1685) is significant. By naming the Royal Society's repository a museum, Grew helped sanction the term for scientific enterprise, endowing it with official status as the title of the Royal Society's collection. While useful for understanding the lineage of 'museum', the *Oxford English Dictionary* is misleading in the case of the word 'repository'. The *OED* states that, denoting 'a place, room or building, in which specimens, curiosities or works of art are collected; a museum', the first use was in 1658, by Edward Phillips.[64] The diarist and Fellow of the Royal Society John Evelyn, however, used the term in his diary entry a full thirteen years earlier, in February 1645. At this time, Evelyn was in Italy, and used 'repository' to describe the collections he visited. [65]

Two decades later, Oldenburg related to Boyle how 'those of ye Society, yt are now in London, doe endeavour to get a good Collection of Naturall and Artificiall Curiosities for ye Societies repository'.[66] Like Evelyn, Oldenburg made good use of the term 'repository'. In a subsequent letter, he described a donation which purchased a 'very handsome Collection of Naturall things for our Repository'.[67] But, despite calling the Society's collection the 'repository' here, on other occasions Oldenburg referred variously to the 'comprehensive Magazeen',[68] the 'Philosophicall Storehouse',[69] the 'collection'[70] or interchangeably to 'repository or Magazeen'.[71]

The fact that Oldenburg used several words to describe the Society's collections illustrates his sense that no one term in the extant vocabulary of collecting was adequate. Oldenburg and his colleagues were trying to distinguish their repository as a new kind of enterprise, distinct from the trivial flaunting of status that defined courtly collecting in pre-modern Europe. The Royal Society's repository was portrayed as a serious natural philosophical project to create a microcosm of nature. Let us consider the language used to describe the objects themselves.

Writing to Richard Norwood in Bermuda, Oldenburg asked him 'to send in what Observables you might meet'.[72] The terms 'rarities', 'specimens' and 'observables' occur throughout his letters to Martin Lister[73] and John Winthrop, the latter from whom he requested 'rarest curiosities'[74] and 'ye chief rarities'.[75] In a subsequent letter he referred to 'ye American Curiosities' that had increased the 'stock of [the Royal Society's] repository'.[76] Oldenburg also asked the diplomat and Fellow of the Royal Society, Sir Robert Southwell, for 'the specimens of all the plants'.[77] It is clear from Oldenburg's correspondence that he used a fairly specific set of terms to describe the Society's collecting project.

Oldenburg was not alone using these terms. The astronomer William Ball, for example, referred to 'all ye curiosities of our own land'.[78] The antiquary Silas Taylor wrote to Oldenburg to tell him he had 'the stones by me for the repository; & what collection I make or can procure you may command'.[79] Richard Stafford, who was briefly the sheriff in Bermuda, wrote to inform Oldenburg that he should receive from Captain Thomas Morlie 'the commandr: of our Magazieene ship such thing as I could at present procure'.[80] John Winthrop wrote frequently about 'the collection of some of the productions of the wildernesse'[81] that he made. The following year he referred to 'the Repository of the Royal Society',[82] and he did so again in a letter describing things he collected and stored at Boston, ready to be shipped back to England.[83]

When John Evelyn attended meetings of the Royal Society he commented with excitement upon the growth of the repository, and used the same set of terms as those of his friends and colleagues. In 1662, he reported that 'at our *Society* ... there was presented for the Repository a piece of Elephants skin, which was about inch-thick'.[84] Six years later he describes another meeting at which there 'were presented ... natural Curiosities'.[85] In 1670, Evelyn 'caled at the Ro: Society ... where were several Curiosities of nature sent us from New England'.[86]

That Oldenburg and his Royal Society colleagues were using a new vocabulary to describe their collecting project is corroborated by the *Philosophical Transactions*. Here, in a public document, Oldenburg made sure that the Royal Society's repository was portrayed in a way that emphasized the seriousness of the enterprise. The introduction to the *Transactions* of 1675 asked 'others in all their *Travels*, by Sea and Land [to] make diligent Researches ... for all Rarities and singularities; that so what is worthy to be acquired ... May (as Arts grow, and as knowledge spreads abroad) be communicated for the benefit of all'.[87] Note the emphasis upon the purpose of the Society's repository.

In 1676, in an extract of a letter concerning New England and North America, the author referred to the Society's 'design of making collections for the History of Nature'.[88] In a later edition, the term 'samples' was used,[89] and 'repository' was frequently the term of choice to describe the Society's collection. In 1670, the *Transactions* printed an extract from one of Winthrop's letters, and referred to 'a very strange and very curiously contrived Fish, sent for the Repository of the R.Society'.[90]

What is the significance of this stable and related set of terms used by the Royal Society Fellows and their colleagues? Bringing the New World under England's intellectual aegis produced a new vocabulary. No matter that the Society fell short of its Baconian ambitions, nor that the reality of its collecting venture was rather more pedestrian and limited than the scrutiny of 'the heavens, the earth [and] the subterranean world' to which Oldenburg aspired. The point is that the Royal Society conceptualized its collecting project as embodying a new approach to nature. In short, a new enterprise required a new vocabulary.

Throughout this book, I have argued that there were two dimensions to man's prelapsarian empire: the recovery of perfect knowledge, which I called the epistemic aspect, and the cultivation of the earth, which I termed the agrarian dimension. The Royal Society, as we have seen, was primarily concerned with the epistemic dimension of man's empire; the recovery of his omniscience. In the following and final chapter, I turn to the thinker who united the epistemic and agrarian dimensions of Adamic empire. This was John Locke.

5 JOHN LOCKE'S LANGUAGE OF EMPIRE

In the past decade, the scholarship on John Locke has recognized the influence of Locke's colonial interests upon his political writings. The 'colonial reading' of the *Two Treatises of Government*, pioneered by James Tully and Barbara Arneil in the early 1990s[1] and more recently developed by Anthony Pagden,[2] Duncan Ivison[3] and David Armitage,[4] has been the major contribution of this strand of scholarship. The majority of these historians have assumed, however, that Locke's interest in the American colonies was generated solely by his professional work for the Earl of Shaftesbury, the Council for Trade and Plantations, the Proprietors of the Carolina Company, and the Board of Trade. They assume that Locke's references to the Americas can be entirely explained by his political defence of colonies as sources of trade, population growth and, above all, property, in the context of a post-Restoration debate about the utility of colonization.

While helpful, this political context does not sufficiently explain Locke's ongoing fascination with colonization and the Americas. In fact this interpretation leaves us with a disjunction between the colonial Locke and Locke the philosopher, Locke the natural scientist and Locke the political economist. I want to suggest that there was another motivation behind Locke's interest in the American colonies and therefore another context in which we should understand his work. Deriving from Locke's concern for the correct use and ends of man's knowledge was his argument that the proper 'Imployment' of men, 'lies in those Enquiries, and in that sort of Knowledge, which is most suited to our natural Capacities and carries in it our greatest interest, i.e. the Condition of our Eternal Estate'.[5] Knowledge, therefore, should help man to extract from the earth the 'Advantages of Ease and Health ... thereby increas[ing] our stock of Conveniences for this Life'.[6] This was an ethos for the improvement of both the earth and the condition of man.

The ethos of 'improvement', the cornerstone of Locke's political philosophy, was a language of empire. Locke articulated the idea of improvement in an explicitly religious context in which improving the earth would return man to his divinely ordained position of dominion over it. Locke's ideas of improvement, formulated in the context of English colonization of the Americas, were part of the tradition of empire that this book has charted throughout the seventeenth century.

Importantly, and unlike many of his colleagues in the Royal Society including Robert Boyle, Locke emphasized that the aim of improvement was not the restoration of Adam's omniscience, but rather the recovery of the earth's Edenic fruitfulness. Locke's epistemological pessimism dictated that natural philosophy could not constitute a 'science' of certainty. Rather, natural philosophical knowledge is possible only insofar as it enables man to reap the conveniences of this life and the fruitfulness of the world. It is because man has the incentive to reap these fruits that he is driven to advance natural knowledge.

Locke's ethos of improvement constituted a fairly stable vocabulary. In a series of terms such as 'husbandry', 'labour', 'industry', 'fruitful', 'employment' and 'usefulness', the vocabulary of improvement embodied the idea that improving the earth is fulfilling the post-lapsarian injunction to render the world fruitful and to glean the 'conveniences for this life'.[7] This was the context in which Locke quoted the line from Paul's First Letter to Timothy that 'God gives us all things richly to enjoy' (1 Timothy 6:17). Locke used this quotation to help justify his argument that man may take what he 'improves' as his property – whether that be in the Americas, Ireland or in England. The language of improvement is thus inherently imperial: we take possession of – have dominion over – what we improve. While the concept of colonialism[8] has greatly improved our understanding of Locke's philosophy, I hope to show that the concept of empire enhances it further.

Locke, the Bible and America

There are several Biblical passages that we can identify as central to Locke's work. Like Bacon, Boyle and the members of the Hartlib Circle and Royal Society before him, Locke often referred to the command in Genesis to 'be fertile and increase, fill the earth and master it; and have an empire over the fish of the sea, the birds of the sky, and all the living things that creep on earth' (Genesis 1:28). In the First Treatise, for example, Locke quotes a later formulation of this idea in Genesis 9:2:

> God, renewing his Charter to *Noah* and his Sons, he gives them Dominion over the *Fowls of the Air*, and *the Fishes of the Sea*, and *the Terrestrial Creatures* ... Wild Beasts and Reptils, the same words that in the Text before us 1. *Gen.* 28 are Translated *every moving thing, that moveth upon the Earth*, which by no means can comprehend man.[9]

This passage was central to Locke's definition of the nature and extent of Adam's original dominion. A second crucial theological idea was that of the Fall. Not only did man decline from perfection but so did the earth itself. Locke refers to Genesis 3:17 which states that 'cursed is the ground for your sake; In toil you shall eat of it'. For Locke the importance of this curse on the earth was twofold. It was marked by the emergence of private property, and it also signified that only through labour, and agrarian labour in particular, will man be able to improve the world and restore its lost fruitfulness.

A third Biblical passage which Locke draws upon as a foundational idea for his work is that of 1 Timothy 6:17: 'God gives us all things richly to enjoy', which Locke quotes in the First Treatise of Government.[10] Central to Locke's work is the idea that God has given the world to man-in-common. In §41 of the First Treatise, for example, Locke develops this idea as an injunction to empire: 'God ... bid Mankind increase and multiply ... to make use of the Food and Rayment, and other Conveniences of Life, the Materials whereof he had so plentifully provided for them ... to promote the great Design of God, *Increase and Multiply*'.[11]

These Biblical passages will be central to this chapter's interpretation of Locke. The theological – and particularly the Adamic – context is largely overlooked by the scholarship which has overwhelmingly focused upon the Second Treatise of Government because of its importance to political philosophy. The First Treatise, however, deserves to be reconsidered by the scholarship. It was in the First Treatise that Locke set forth the theory of Adamic empire over the earth which was so vital to the rest of his work.

The pioneer of the colonial strand of Lockean historiography, Barbara Arneil, placed Locke in the seventeenth-century debate about colonization. In the context of misgivings by men such as Roger Coke,[12] defenders of plantations such as Josiah Child and Charles Davenant argued that 'large scale settlement and cultivation of new lands' could create wealth and that this was an answer to England's fiscal woes.[13] The fifth chapter of the *Two Treatises of Government*, Arneil argues, was a 'vigorous defense of England's colonial activities in the new world'.[14] Arneil shows that Locke's Second Treatise was 'almost exclusively' concerned with labour in terms of crop-growing, agrarian activity,[15] which was precisely the type of agriculture pursued in Carolina.

In discussing Locke's agrarian language, however, Arneil uses only the context of economic debates about colonies to illuminate Locke's ideas. The terms that Arneil quotes from the *Two Treatises*, such as 'plough, sow and reap' and 'improvement, tillage or husbandry' and 'Pasturage, Tillage, or Planting'[16] were, as I will show in the next section, derived from Locke's Biblical argument about man's duty to restore the world to its original fruitfulness and command it as his empire. While it is a generally secular reading, Arneil's argument is persuasive; Locke's political philosophy was forged on the back of his colonial interests.

In a recent article, David Armitage advanced this thesis considerably by providing new evidence that the revision of the *Fundamental Constitutions of Carolina* was Locke's direct concern when composing his chapter on property in the *Two Treatises*. Armitage demonstrates that Locke was involved in the third revision of the *Fundamental Constitutions of Carolina* in the summer of 1682.

This fact corresponded to a shift in Locke's theory of property, from 'a broadly Grotian account of the process by which the primal positive community in the world had given way to the regime of exclusive private property'[17] to one based upon the 'divine command' to cultivate undeveloped land.[18]

Armitage shows that 'Locke's argument from divine command to cultivate those "*great Tracts*" of un-appropriated land became the classic theoretical expression of the agriculturalist argument'[19] for European claims over American land. This is no doubt the case, but it is curious that despite referring here to Locke's use of a Biblical argument for the cultivation of the world, Armitage does not explore the significance of Locke's theological argument. Rather, he focuses upon how Locke's agriculturalist argument replaced earlier arguments for possessing land based upon conquest and religious conversion of native peoples. As I will show, however, Locke's references to the divine command to develop the earth and render it fruitful indicate that he was employing a theory of empire. Such Biblical language dominates the *Two Treatises of Government*.

In *An Approach to Political Philosophy: Locke in Contexts*, James Tully places Locke's *Two Treatises* in the context of existing justifications for sovereignty in the New World, and reveals the two conceptual acts Locke performed by placing the Americas in the state of nature. First, Amerindians' political organization could be disregarded in favour of the model of individual self-government; second, their system of property could be ignored in favour of an individual labour-based theory.[20] This is important for our understanding that Locke's chapter on property was an attempt to justify English claims to Amerindian land.

Tully's reading of the *Two Treatises* in *Locke in Contexts*, however, is predominantly secular. This fact marks a contrast with Tully's first book, *A Discourse on Property*, which deals extensively with the theological dimension of Locke's work, recognizing that 'if there is one leitmotiv which unites Locke's works it is surely a philosophy of religious praxis'.[21] In this book, Tully shows that Locke was arguing that property is not private dominion but a right in common with all mankind. In short, this right in common was the expression of the dominion God gave mankind in Genesis: 'the Dominion of the whole Species of Mankind, over the Inferior Species of Creatures'.[22] Rights, therefore, are 'use-rights' which permit the collective exploitation of the earth. It is 'a Right, to make use of the Food and Rayment, and other Conveniences of Life, the Materials whereof he had so plentifully provided for them'.[23]

Man is able to grasp this understanding of the normative relationship between himself and the world not only through the Bible, but also because it is evident in natural law. Dealing with the intersection of scripture and natural law in Chapter V of the Second Treatise, Tully writes that 'in the first eight lines Locke sets out the two initial conditions which partially define man's natural state. Scripture reveals that the world is a gift, given by God to mankind in common. Natural reason teaches that each man has a right to the things which nature affords for his subsistence.'[24] Thus, according to both scripture and natural law, God's purpose for man was that he use the world for his preservation.

In dealing with the theological content of Locke's work, Tully developed the attention to Locke's religious views pioneered by John Dunn and developed recently by Jeremy Waldron[25] and Peter Harrison. Harrison points out the significance of Biblical justifications for colonization in seventeenth-century England, focusing on the injunction in Genesis 1:28 to 'fill the earth and subdue it', and recognizing its centrality to Locke's theory of property.[26]

It is now well established that Locke formulated his theory of property in the context of his involvement with the Proprietors of the Carolina Company. In 1668, only two years after he first met his patron Anthony Ashley Cooper, Locke became secretary to the Lords Proprietors of Carolina. He served as physician and secretary to Lord Ashley, who later became the first Earl of Shaftesbury. Locke managed his colonial affairs and lived in the Shaftesbury household in London. Here Locke helped to draft the *Fundamental Constitutions of Carolina* in 1669. In 1672, Locke purchased the first of two sets of shares he was to own in colonial companies. He invested £600 in the Royal Africa Company, which traded along the West Coast of Africa and provided slaves for the colonies in the West Indies.[27] That same year, Locke became one of eleven Bahamas Adventurers. This was a project with his employer and five of the other Carolina Proprietors.[28] As Wayne Glausser has shown, Locke 'initially invested one hundred pounds; before long he doubled his share by taking over the one hundred-pound investment of his friend John Mapletoft'.[29]

Locke became Secretary to the Council of Trade and Foreign Plantations in 1673 and served for two years. When the Board of Trade was established in 1696, Locke was one of King William's eight Commissioners. In this capacity, Locke's task was to oversee the colonies 'with regard to the administration of the government and justice in those places as in relation to the commerce thereof'.[30] In addition, Locke corresponded with people about the events and natural philosophy of the Americas. Locke's correspondents in the Americas included the Revd James Blair in Virginia, Isaac Rush, who was Shaftesbury's deputy in New Providence Island in the Bahamas, Joseph West, briefly the Governor of Carolina and an associate of Shaftesbury, and Sir Peter Colleton, one of the Carolina Proprietors as well as the owner of one of the largest slave plantations in Barbados.

Locke's Philosophy and the Language of Improvement

Here then is a large field for knowledge proper for the use and advantage of men in this world, viz., to find out new inventions of dispatch to shorten or ease our labours, or applying sagaciously together several agents and patients to procure new and ben-

eficial productions whereby our stock of riches (ie things useful for the conveniences
of our life) may be increased or better preserved.

John Locke, 'Understanding' (1677)[31]

The idea of improvement was the abiding core of Locke's philosophy. Improving
the world in order to reap the fruits that God has endowed for mankind was a
central tenet of Locke's epistemology, his theology, his natural philosophy and
his political economy. 'Improvement' was a theory of empire; it embodied the
idea of the restoration of man's dominion over the earth.

In the quotation above, taken from his essay on 'Understanding' written
in 1677, Locke explains his view of the potential for, and correct use of, man's
knowledge. The human mind 'finds itself lost in the vast extent of space, and the
least particle of matter puzzles it with an inconceivable divisibility'.[32] The kind
of knowledge man can have is 'sufficient if we will but confine it within those
purposes and direct it to those ends which the constitution of our nature and the
circumstances of our being point out to us'.[33]

Locke's description of the proper field of enquiry for mankind presents us
with an answer to two related questions. Why include Locke in a book about
natural philosophy? What, moreover, was Locke's conception of natural philosophy? If we were to apply the categories of early modern enquiry, we could argue
that Locke's interest in the New World encompassed the fields of natural history, moral as well as natural philosophical information. This is one legitimate
approach. To divide up Locke's interest in the natural world and the New World
like this, however, would limit our understanding of his work and potentially
lead to the false assumption that these categories were themselves stable divisions
of the universe. As Katharine Park and Lorraine Daston point out, the early modern period witnessed the 'merging of natural history with natural philosophy'.[34]

While it is vital to recognize the limitations inherent in using the term 'natural philosophy' as a kind of methodological umbrella, in the case of Locke it
is both justifiable and helpful. The most compelling reason to consider Locke's
views on the natural world under this general, albeit imperfect, concept is that
Locke's own conception of knowledge was broadly natural philosophical in the
tradition of the new philosophy which this book has charted. As Mark Goldie
has pointed out, Locke in his explicit epistemological statements made Baconian
and Boylean appeals to 'avoid abstract metaphysical speculations, and to seek
instead the acquisition of useful, "experimental" knowledge'.[35] Moreover, in his
essay on 'Understanding', Locke did not subscribe to any categorical distinction
between natural history and natural philosophy. On the contrary, in a Baconian
moment, Locke included 'the history of nature and an enquiry into the qualities
of the things in this mansion of the universe' and the 'material causes and effects

of things in their power ... arts and inventions, engines and utensils' under the one rubric as the proper subject matter for man's knowledge.[36]

Locke's conception of knowledge, then, was broadly natural philosophical, practical and in the Baconian tradition. The 'large field for knowledge proper for the use and advantage of men in this world' was intended to 'procure new and beneficial productions whereby our stock of riches (i.e. things useful for the conveniences of our life) may be increased or better preserved'.[37] Locke's description of the purpose and subject matter of knowledge was not, however, accompanied by an epistemological confidence about its outcomes. He doubted the possibility of man's omniscience, because 'our minds are not made as large as truth nor suited to the whole extent of things amongst those that come within its ken'.[38]

Throughout his work, Locke exhibited what Michael Ayers has called an 'epistemological pessimism' as to the possibility of a proper science of nature based on the essences of things.[39] In Chapter XII of Book 4 of the *Essay Concerning Human Understanding*, Locke explains his epistemological doubts: 'This *way* of getting, and *improving our Knowledge in Substances only by Experience* and History, which is all that the weakness of our Faculties in this State of *Mediocrity*, which we are in this World, can attain to, makes me suspect, that natural Philosophy is not capable of being a Science.'[40] Similarly, in *Mr. Locke's Reply to the Right Reverend Lord Bishop of Worcester*, who critiqued his *Essay Concerning Human Understanding*, Locke explained that it was 'evident that our Knowledge cannot exceed our Ideas, where they are only imperfect, confused or obscure'.[41]

Locke's epistemological doubt is at odds with the fervent optimism about the possibility of recovering universal knowledge maintained by his Royal Society colleagues such as Robert Boyle. Boyle held dear the project of recovering Adam's epistemic empire and, as we saw in the previous chapter, this ideal was shared by members of the early Royal Society, many of whom, like Locke, cultivated related interests in both natural philosophy and the Atlantic colonies. Moreover, Locke admired Boyle and actively oversaw the posthumous publication of Boyle's *Generall History of the Air* in 1692, upon which he made corrections and wrote the Advertisement to the Reader.[42]

Although he did not share many of his friends' optimism, Locke did share their belief in the significance of the Fall. The most important aspect of the Fall for Locke was not man's corruption but that of the earth. When we explore Locke's philosophy we find it imbued with the sense of recovering the fruitfulness of the fallen earth for the benefit and enjoyment of mankind. The purpose of knowledge is to endow man with the faculties to 'apply [parts of the universe] to our uses and make them subservient to the conveniences of our life, as proper to fill our hearts and mouths with praises of [God's] bounty'.[43] This is an idea of *empire*; it is the ideal of fulfilling God's plan by returning man to his intended dominion over nature.

There are many references to Adam's original empire and Fall, as well as the earth's attendant corruption, in Locke's work. In his essay 'Homo ante et post Lapsum' ('Man before and after the Fall'), for example, Locke states that before the Fall 'Man was ... put into a possession of the whole world, where in the full use of the creatures, there was scarce room for any irregular desires'.[44] Locke then cites Genesis 3:22 in which man was 'thrust from the tree of life'.[45] The Fall, Locke explains, was the cause not only of man's corruption, but also of a 'curse on the earth'[46] which precipitated the emergence of private property and consequent inequalities between men: 'When private possessions and labour, which now the curse on the earth made necessary, by degrees made a distinction of conditions, it gave room for covetousness, pride and ambition, which by fashion and example spread the corruption which has so prevailed over mankind.'[47]

Locke also drew upon this Adamic idea in the Second Treatise of Government. When arguing for parents' duties towards educating their children in the chapter on paternal power, Locke described precisely this notion of Adam's original perfection and man's post-Lapsarian corruption. '*Adam* was created a perfect Man, his Body and Mind in full possession of their Strength and Reason, and was so capable from the first Instant of his being to provide for his own Support and Preservation and govern his Actions according to the Dictates of the Law and Reason which God had implanted in him. From him the World is peopled with his Descendants, who are all born Infants, weak and helpless, without Knowledge or Understanding.'[48]

Locke's references to the Adamic idea and its implications for the re-creation of man's dominion are a powerful theme throughout his work. In the *Essay Concerning Human Understanding*, Locke uses Adam to illustrate his epistemological argument that words signify ideas rather than real essences. In the tale of Adam and Lamech, Locke supposes that '*Adam* in the State of a grown Man, [had] a good Understanding, but in a strange Country, with all Things new, and unknown about him; and no other Faculties, to attain the Knowledge of them, but what one of this Age has now'.[49] Locke tells the story of Adam's suspicion that his friend Lamech is troubled. Adam uses two new words, only to discover that neither describes the essence of Lamech's troubles. So, Locke argues that Adam's names referred to his own complex ideas instead.[50]

Locke draws upon the idea of Adam and his children in a later section in the *Essay* to describe Adam's naming of substances, in particular a 'glittering substance' which is bright yellow.[51] This time Adam appears as the first natural philosopher as he 'knocks it and beats it with Flints, to see what was discoverable in the inside'.[52] As Hans Aarslef has pointed out, part of Locke's position that there are no innate ideas is the argument that mankind also possesses no vestiges of an original, perfect Adamic language.[53] Here in the *Essay* Locke insists upon Adam's ordinary humanity in regard to language.[54] Adam was no more able to represent the perfect essence of things in words than men in Locke's own century were. For Locke, it was not Adam's epistemic empire that is important for mankind; it was the original fruitfulness of the earth which man must endeavour to restore. It is surprising that Locke's theological argument is rarely used to help understand his philosophy of improving the world.

Locke articulated the idea of recovering the earth's fruitfulness in a vocabulary of 'improvement' which included terms such as 'innovations', 'advancement', 'dominion', 'subsistence', 'useful arts', 'employments', 'labour', 'trade', 'exercise', 'industry', 'plenty' and 'conveniences of life'. The vocabulary also includes a series of negative terms against which the above are contrasted, such as 'idleness', 'waste' and 'ignorance'.

The primary place where we see the importance of the language of improvement is in Locke's epistemology. Chapter XII of Book 4 of the *Essay Concerning Human Understanding*, for example, is titled 'Of the Improvement of our Knowledge'. Just after explaining his epistemological doubts as to the possibility of a *science* of nature, Locke reassures his audience that we are capable of drawing 'the Advantages of Ease and Health' which will 'increase our stock of Conveniences for this Life'.[55] Locke echoed these ideas in his *Reply to the Bishop of Worcester*: 'It becomes our Industry to employ it self, for the Improvement of the Knowledge, and adding to the Stock of Discoveries left us by our inquisitive and thinking Predecessors.'[56] Locke directed the bishop to read his 'Chapter of the improvement of our Knowledge' in his *Essay Concerning Understanding* which would 'shew how necessary it is for the enlarging of our Knowledge'.[57]

The Americas provide a useful comparison when Locke discusses the centrality of knowledge in improving the world. In the chapter of the *Essay* cited above, Locke uses the language of improvement to contrast the natural philosophical discoveries in England to the lack of such innovations in America. Locke's vocabulary is striking so the passage deserves to be quoted at length:

> Of what Consequence the discovery of one natural Body, and its Properties may be to humane Life, the whole great Continent of *America* is a convincing instance: whose Ignorance is useful Arts, and want of the greatest part of the Conveniences of Life, in a Country that abounded with all sorts of natural Plenty, I think, may be attributed to their Ignorance, of what was to be found in a very ordinary despicable Stone, I mean the Mineral of *Iron*. And whatever we think of our Parts or Improvements in this part of the World ... it will appear past doubt, that were the use of *Iron* lost among us, we should in a few Ages be unavoidably reduced to the Wants and Ignorance of the ancient savage *Americans*.[58]

In his essay 'Some Thoughts Concerning Reading and Study for a Gentleman', Locke places 'books of Travel' on the list of his recommendations.[59] Locke reveals his knowledge of such books by referring to English authors such as 'Hakluyt and Purchas' and 'Sandys, Roe, Brown, Gage and Dampier' as well as authors in

French and Italian, most of whose books he owned.[60] Travel books exhibit 'a very good mixture of delight and usefulness'.[61]

Another comparison between the Amerindians and the English occurs in Locke's essay 'On Study' which he wrote in 1676. Here the vocabulary of improvement places the Amerindians in an anthropologically primordial moment: 'Perhaps without books we should be as ignorant as the Indians, whose minds are as ill-clad as their bodies. But I think it is an idle and useless thing to make it one's business to study what have been other men's sentiments ... knowing of these opinions ... may serve to instruct is in the vanity and ignorance of mankind.'[62]

The best illustration of Locke's idea that the purpose of knowledge is to aid the restoration of man's empire over the material earth by rendering it fruitful is contained in the First Treatise of Government. The theme of the text is the type of dominion or empire that Adam possessed. Arguing against Robert Filmer, Locke's position is that Adam and his male descendants were *not* given the arbitrary power over life and death. Rather, they were given a very different kind of power. This was the power to render the world fruitful, and thus make it their property. Citing Genesis 9:2, Locke writes, 'God, renewing his Charter to *Noah* and his Sons, he gives them Dominion over the *Fowls of the Air*, and *the Fishes of the Sea*, and *the Terrestrial Creatures* ... Wild Beasts and Reptils, the same words that in the Text before us 1. *Gen.* 28 are Translated *every moving thing, that moveth upon the Earth*, which by no means can comprehend man.'[63]

Locke's speech here, as Peter Laslett noted long ago, was a critique of Filmer's argument for absolutism derived from God's command to Adam. This is of course true, but here Locke was not only making a political argument against absolutism. He was explaining the *theological* foundations of an entire plan for mankind; God has given the entire material globe to man, whose job it is to reap the earth's goods and ascend to the position of dominion that he originally commanded. In employing the concept of empire in the *Two Treatises*, Locke was performing a political speech act; he was bringing the Biblical ideal of man's plenary dominion over nature into a political tradition.

Locke explained God's plan for mankind using the language of improvement. God 'furnished the World with things fit for Food and Rayment and other Necessaries of Life, Subservient to his design'. Man should 'use those things, which were serviceable for his Subsistence, and given him as means of his *Preservation* ... Man had a right to a use of the Creatures, by the Will and Grant of God.'[64]

Clearly, Locke was outlining a theory of empire. God set 'mankind above the other kinds of Creatures, in this habitable Earth of ours. Tis nothing but the giving to Man, the whole Species of Man, as the chief Inhabitant, who is the Image of his Maker, the Dominion over the other Creatures.'[65] Locke uses the term 'empire', regardless of its synonyms, no less than twenty-three times in the Two Treatises.

Some caution and careful analysis is needed in our attempt to understand how Locke used the terms 'empire' and 'dominion'. It is true that he often used the two words interchangeably, for example when he argued that, as the elder child Cain did not have 'Dominion' over Abel,[66] and when he described the power of the God-as-maker over man, his creation: 'A dependent intelligent being is under the power and direction and dominion on whom he depends and must be for the ends appointed him by that superior being.'[67]

Both 'empire' and 'dominion', therefore, could be used interchangeably as metaphors for political authority. In these abstract political usages, empire has no necessary relationship to property. Two useful illustrations include, first, the case in which 'Empire' in fact falls to the Younger Brother,[68] and, second, the fact that in the time of the patriarchs there was 'no knowledge, no thought, that Birth-right gave Rule or Empire'.[69] A third example is Locke's reference to Filmer's idea that 'Empire belongs to the Eldest Son and his Heirs'.[70] Here empire clearly means straightforward political sovereignty, rather than property.

My point is not to deny that the terms empire and dominion functioned as metaphors for political authority, but rather to suggest that Locke used these terms *both* as metaphors of power *and* to refer to his theory of empire. Although this is obviously a complex point and does not paint Locke's language in clearly defined colours, I do not think it is implausible to suggest that Locke used the words as metaphors as well as to refer to his theory of empire. After all, we are all are perfectly capable of making the same word function at some moments as an abstract metaphor and at others as a specific label. We might, in our contemporary parlance, use the term 'God' in this manner: in some contexts 'God' stands for all transcendental power, but in other contexts God denotes one specific deity. This analogy, I think, best describes how Locke used the terms 'empire' and 'dominion'. In some contexts, they were metaphors of political power, but in other contexts they signalled an underlying conception of man's plenary empire over the earth.

James Tully has also pointed out Locke's linguistic equivocity, referring to the terms property and rights. Locke

> is clearly aware of the equivocity and it seems to be simply a continuation of the equivocity of similar Latin terms such as *ius* and *dominium*. Equivocity is normally a linguistic signal that two items are related in some way; a relation which might go unnoticed if two different terms were used. In this respect, equivocity is different from ambiguity, where two items bear the same name but do not stand in any relation one to another.[71]

I would add a further point here. Locke's linguistic equivocity with the terms property, empire, dominion and right, does not mean that every use of those terms is equally significant. Some linguistic instances are more instructive and revealing than others, and we come to determine the most significant linguistic moments from an assessment of their context. It is when Locke uses a vocabulary of empire to refer to the theory of man's Adamic dominion over the world that his language is most revealing. This is because in using this language Locke was bringing the theological idea of Adam's empire into the context of political thought.

America and the Improvement of Natural Philosophy

The idea of improving the world was the core of Locke's natural philosophy and he enthusiastically sought natural information from the Americas. Throughout his life Locke was intensely interested in, and assured of, the benefits of useful knowledge. Following the Baconian tradition, Locke maintained that the purpose of natural philosophy was to attain that knowledge which would aid 'the business of men being happy in this world and by the enjoyment of the things in nature subservient to life, health, ease and pleasure'.[72] Natural philosophy was a field of knowledge that would aid the improvement of the world, and should itself be advanced. Man must search for that natural knowledge which 'may be attained by moderate industry and improved to our infinite advantage'.[73]

Locke's emphasis upon the utility of natural philosophy, and upon its role in the improvement of man's estate was partly the result of his belief that natural philosophy cannot reach absolute certainty, and thus man must instead use it 'as an enlargement of our Minds towards a truer and fuller comprehension of the intellectual World, to which we are led both by Reason and Revelation'.[74] Natural philosophy would never become a science,[75] but Locke assured his readers of *Some Thoughts Concerning Education*, that 'I would not deter any one from the study of Nature, because there are very many things in it, that are convenient and necessary to be known to a Gentleman: And a great many other, that will abundantly reward the Pains of the Curious with Delight and Advantage'.[76] Here Locke's language of improvement is marked: terms like 'advantage', 'convenient' and 'necessary' came to characterize Locke's work.

Such 'advantages' and 'improvements' within natural philosophy were often of a practical nature. 'Rational Experiments and Observations' were much more useful than 'barely speculative Systems'.[77] The practical arts were inherently part of natural philosophy. When advising gentlemen about education, Locke insisted that '*Husbandry, Planting, Gardening*, and the like, may be fit for a Gentleman, when he has a little acquainted himself with some of the Systems of the *Natural Philosophy* in Fashion'.[78] Locke was fascinated by the practical nature of natural philosophy. Between 1666 and 1703 he kept extensive and detailed records of the weather, noting the rainfall.[79] Some of these he published. The Royal Society's *Philosophical Transactions*, for example, published 'A Register of the Weather for the year 1692, kept at Oates in Essex'.[80] This was part of the ideal of collating a practical, extensive bank of natural knowledge.

Weather records aside, much of Locke's interest in natural philosophy was provided by the Americas. In fact, the Americas provided information about four aspects of natural philosophy that were of interest to Locke. The first was a general concern with the natural environment, climate, flora and fauna. While he was Secretary to the Lords Proprietors of Carolina from 1668 to 1675, Locke corresponded with several men in the colony. In 1690, for example, Locke made extracts from a series of letters he received from John Stewart, concerning Carolina's natural environment. Stewart's descriptions were arranged under various headings, such as 'air', 'rivers', 'trees', 'shrubs', 'diseases', fosilia', 'quadrupides', 'commodities' and 'soyle'.[81] Under the heading 'trees', Locke noted Stewart's list of 'Cedars, Cypresse, Oakes, pines of severall kindes, Bay Dogwood, Hickery or wild walnut'.[82] The heading 'fish' included 'oysters, muscles, clams, crabs all in plenty'; 'birds' included 'Wild Turkey large and fat. Ducks and Goose. Turtle doves in winter'; 'Serpents' included 'vipers, Rattlesnake, Blacksnake, chicken snake, Lizards [and] vipers'.[83] The fact that Locke transcribed these natural phenomena in such detail and taxonomy is testament to his fascination with the information about the natural world to be discovered in America.

Locke was proactive in asking his correspondents in the Americas to send him details of the flora and fauna, as well as the weather. Among his manuscript papers on Virginia, for example, is a series of 'Queries' about the 'land, people, constitution and revenue of Virginia'.[84] Section 5 was devoted to questions about 'The Navigation'.[85] While he was Secretary to the Council of Trade, in May 1675, Locke sent Henry Oldenburg 'an account I lately received from New: Providence one of the Bahama Islands concerning fish there'.[86] Referring to the man who had sent him the account of the fish, Locke said: 'I expect to receive from him in answer to some Queries I lately sent him by a ship bound thither.'[87] That same year, Locke's letter was published in an edition of the Royal Society's *Philosophical Transactions*, under the heading 'Extract of a Letter on Poisonous Fish'.[88]

Locke received specimens as well as written reports. Dr Henry Woodward wrote to Locke from Westo (in present-day Georgia), telling him about the 'hearbs and roots' of the native Americans 'which they impart onely to the next akin. Had I bin upp in the maine I should have sent some now, but shall by the next opportunity'.[89] The Deputy Governor of Barbados, Sir Peter Colleton, wrote to Locke discussing many aspects of the colony of Barbados, such as its trade, as well as describing various native plants and their uses in the Amerindians' medicine. Colleton sent Locke some China root (what we know today as Cinchona bark). The bark is the source of quinine, which was used to treat malaria. Writing on 12 August 1673, Colleton tells Locke that 'by the last Fleet I sent you a parcel of Carolina China Root, which was directed to Colonel Thornburgh for you, by this I send you a Jar of this countrey Tar, which I think is Oyl of Bitumen of whose sanative quality some here talke wonders, I have Known the Oyl of it helpe the sciatica, and it with white lilly root hath cured the Glanders in severall of my horses, I also send you a pott of Tarara root, which cures the wounds made by the Indians poisoned Arrows, which was first discovered by Major Walker.'[90] Tarara root is arrowroot, and its rhizome is an antidote to poison.[91]

The second type of natural philosophical knowledge that interested Locke was the medicinal qualities of plants and animals. As a qualified physician, Locke was naturally interested in medicines, as his manuscripts attest. The fact that there existed in the Americas a whole new botanical world of potentially medicinal plants excited him. In the above letter, Colleton went on to describe how the medicinal properties of the tarara root were used: 'An Indian that had accidentally prick'd his Thumb with an arrow ... and having none of this root gave himself over for dead, and his hand swell'd extreamly, Major Walker[92] being with him found amongst his things a small piece of the root at the sight of which the Indian rejoiced ... and applying some of it to the wound ... by the fresh Juce of that root quite cured his Thumb in a very short time.'[93] More medicinal information was provided by Richard Lilburne, about whom little is known, apart from the fact that he lived in Providence Island in the Bahamas. In August 1675, Lilburne wrote to Locke and described the medicinal properties of the oils extracted from various animals: 'Here is Snakes oyle Guana's and Aligater's which last I hear is an admirable remedie for the Gout.'[94]

Apart from medicinal properties, plants were often useful in other ways. Some natural products had qualities that made them useful for trade. In October 1673, Sir Peter Colleton wrote to Locke and described Brasiletto wood, which was a type of dye-wood. Like Locke, Colleton was one of the eleven Bahamas Adventurers, and in this letter he discussed the potential trade for that company. Brasiletto wood was a possibility if 'you Inquire amongst the Dyers whether Brasiletto be of absolute necessity for the dying of any couller, or whether onely to helpe when Logwood is deare, for it be onely used in that case ... the price is not like to rise and thus little profit would be made'.[95]

In the same letter, Colleton described the natural environment of Barbados. Once again, the information is bound up with ideas of utility, this time in reference to agricultural improvement: 'The country is extream healthy and pleasant, and the understanding planters say its very fertill, but better further up than where they are Setled, which is soe near the barren sands of the Sea shore, I am very sure that if we overcome the want of Victuall, all the English planted northward will come into us for in new England the greatest [part] of the summer labour of the husbandman, is spent to p[rovide] fodder ... the cattle of Carolina were beef ... and will never need to be fothered which advantage added to our being able to produce many commodityes that they cannot.'[96]

Locke collected a significant amount of knowledge about the natural environment of Carolina to the extent that some men, including Peter Colleton, regarded him as an authority on the colony's natural environment. We know this because in 1671 Colleton wrote to Locke and explained that a bookseller, Mr Ogilby, was printing 'a relation of the West Indieas' and asked Colleton to 'gett a map of Carolina'.[97] Colleton asked Locke to source the map from Lord Ashley, and asked Locke 'if you would due us the favour to draw a discourse to bee Added to this map in the nature of a description such as might invite people [in order to] ... very much conduce to our speedy settlement'.[98] Locke was evidently seen as the right man for the job. Although a very favourable description of Carolina's natural environment was eventually published in Ogilby's *English Atlas*, it does not appear that Locke was its author.[99]

Another area about which Locke requested and received significant amounts of natural philosophy was the Bahamas. Richard Lilburne's letter to Locke in 1674 reveals that Locke had sent him various enquiries about the natural world of New Providence Island:

> I have not met with any rarities worth your acceptance though I have been diligent in inquireing after them, of those which I have heard of one seems strange to me: the fish which are here are many of them poisonous bringing a great pain ... their joints which eat them and continues soe some short time and at last with 2 or 3 dayes itching the pain is rub'd off.[100]

Lilburne continues, 'I think you spoke to me of some oyle of souldiers [crabs] and I have indeavoured to procure it but his being the time of their spawning non can be drawn from them: when the opportunity presents it self I shall be ready to serve you in that or any thing el<se that you> will command.'[101] This is a good example of the way that information about natural philosophy encouraged Locke to make his own inquiries into the Americas.

Lilburne evidently managed to observe some of the crabs and, though he doubted his own talents in natural philosophy, he wrote to Locke again in August 1675 and reported:

> according to their growth they provide themselves still with shells suitable ... of which they take possession for their own use and draw about with them till by their growth they find themselves too much pent up and then they make inquiry for a new habitation ... the oyle of Soldiers is much esteem'd here and difficult to be got they reckon it here a very sovereign thing for any aches and pains either in the limbs or joynts.[102]

Lilburne went on to say that 'there is one who I imploy'd at Andrews Island[103] to get some Souldiers and try them in the Sun who hath sent down word he hath procur'd some which if it come before the Ship goe away I shall send it'.[104] Lilburne concluded his letter with 'your particular queries I have answe'd in a piece of paper inclosed'.[105]

Enclosed with Lilburne's letter were 'Answers to Queries propos'd according to the best account, and observations which I could meet with here'. The observations are numbered and must refer to numbers that Locke had attached to his own questions. They relate primarily to the nature of poisonous fish. The second point, for example, explains that 'the poisonous which I could get account of is Rock-fish, Parcudas, Amber-fish, Hogg fish, Snappers etc'.[106] Points 3 and 4 explain other types of poisonous fish. In point 7, Lilburne explains that 'the women and children and weakest constitutions are soonest infected' and in point 8 he describes the symptoms, which include 'tingling in the nose ... a violent itching in all parts'.[107] There are thirteen points in total.

A number of Locke's correspondents commented upon the poisonous fish of the Caribbean. Isaac Rush, who also lived in New Providence, described the effects of poisonous fish upon humans, as well as the medicinal qualities of other animals. In 1675, Rush told Locke that he 'sent a letter in which I gave the best account I could of the nature of our poison fish ... next I will indeavour to procure some Gwyana and some snakes oyle, which is reported to be of greate use: I have directed a small box to thee with some shells and other trifles in it'.[108] Locke greatly appreciated medical information because it was useful knowledge; it contributed to the improvement of natural philosophy.

The third type of natural philosophical information that Locke was eager to receive was that related to the Amerindians themselves. Dr Henry Woodward, in Westo, wrote to Locke in November 1675 and reported answers 'concerning the religion and worship, Orginall, and customes of our natives, especially among the Port Royall Indians' that Locke had obviously asked him to glean.[109] Woodward reported that the Port Royall Indians, who lived on the coast of South Carolina, south of Charleston, 'acknowledge the sun to bee the immediate cause of the groth and increase of all things whom likewise they suppose to be the cause of all deseases'.[110]

Woodward also provided information about a subject within natural philosophy which interested Locke as well as his Royal Society colleagues (as we saw in the previous chapter). This was the issue of whether the Noachian flood reached America. Locke owned Thomas Burnet's *Theory of the Earth* (1684) as well as his *Reflections upon the Theory of the Earth* (1699) and his *Archaeologiae Philosophicae* (1692).[111] The Amerindians, Woodward recounted, 'have some notions of the deluge, and say that two onely were saved in a cave, who after the flood found a red bird dead: the which as they pulled of his feathers between their fingers they blew them from them of which came Indians ... and they say these two knew the waters to bee dried up by the singing of the said red bird'.[112]

Joseph West, stationed in Charlestown, Carolina, wrote to Locke in 1676 and presented him with 'the best account I can get Concerning the Natives here'.[113] Locke had evidently also asked for the hides of some deer, because West replies: 'Sir did I know where to get any dressed Deare Skins worth Your acceptance, I should Gladly present them to you.'[114] His letter described the condition of the Indians in Carolina and the nature of their trade.

Locke's interest in what we would call the anthropology of indigenous Americans also manifested itself in the notes he took from books that he read. Using his method of common-placing, Locke took notes on Gabriel Sagard Theodat's *le grand voyage du pays des Hurons*. The heading 'Peccatum' for example, pertained to 'a people upon the river of Amons whose religion forbids them to eat any other flesh only that of men'. He then records that 'there is another people on the same river where the old people only labour and the young men and women have no thing to doe but to please them selves & get children'.[115]

The fourth aspect to Locke's interest in the natural world of the Americas centred upon his membership of the Royal Society and friendship with some of its most illustrious members, including Robert Boyle and Isaac Newton. After Boyle's death in January 1691, Locke edited and published his *Generall History of the Air*. Of all Boyle's works, *The Generall History of the Air* was possibly the most heavily reliant upon Boyle's extensive interviews with travellers to what Boyle termed 'outlandish places', many of which were in the Americas. Observations about salts, for example, were derived from 'an acquaintance that lived long in parts of Americas',[116] and Boyle's work on the effects of air upon iron were derived from 'a learned observer who practis'd Physick in one of the most Southern of the English Colonies'.[117] Locke no doubt enjoyed reading and editing the many natural philosophical references to the natural world of the Americas.

Locke wrote the 'Advertisement' to the Reader, and made it clear that Boyle's work was part of a general project of the improvement of natural philosophy. He described Boyle's 'noble, and always busy Designs for the Advancement of Knowledg, and the Benefit of Mankind'.[118] Locke then stated explicitly the role of natural philosophy in 'improving' the world. 'We must not expect to find in every Age a Man able and ready to lay out so much Cost, Pains and Skill, frankly, for the Improvement of Natural Philosophy, and the Information of the World, as he has done.'[119]

The Political Economy of Empire in the Atlantic Colonies

We have seen how Locke's imperial language of improvement and attendant interest in the Americas shaped his natural philosophy. Let us now focus upon Locke's political economy. The idea of improving the world and returning it to man's dominion was central to Locke's dealings with England's colonies in the

Atlantic. Locke's political economy consisted of a series of practical schemes which involved educating and employing the poor, the effective regulation of trade, a general naturalization of immigrants, and ensuring full population where possible. Political economy was designed to minimize idleness and in favour of industry, regulation and employment, all of which were key terms in Locke's language of improvement.

The centrality of the idea of improvement to Locke's political economy is no better illustrated than in the *Two Treatises of Government.* The colonial reading of this text claims a special place for America in the chapter on property. It is highly likely, as Armitage has shown, that Locke's revision of the *Fundamental Constitutions of Carolina* in the summer of 1682 was the immediate context for Locke's composition of Chapter V.

The idea of improvement, however, was fostered not only by Locke's colonial interests but also by his theory of man's plenary empire over nature. Mixing one's labour with the land was fulfilling God's divine plan that man should improve the earth and restore his dominion over it. Locke's labour theory of property, therefore, was also a theory of empire, and the American colonies were the ideal place to put this idea into practice.

The most powerful evidence for the theological and imperial nature of the *Two Treatises* is the fact that whenever Locke articulates his point about improving the earth and mixing one's labour with the land, he does so by making reference to God's plan for mankind. In Chapter IV of the First Treatise, when discussing Adam's sovereignty, Locke outlines the point to which he returns frequently: '*Be Fruitful, and Multiply, and Replenish the Earth*, says God, in this Blessing.'[120] Arguing against absolute monarchy, Locke again speaks of the political economy of improvement in a theological context:

> For how much Absolute Monarchy helps to fulfil this great and primary Blessing of God Almighty *Be fruitful and multiply and replenish the Earth,* which contains in it the improvement too of Arts and Sciences, and the conveniences of Life, may be seen in those large and rich Countries, which are happy under the *Turkish* Government, where are not now to be found ... not 1/100 of the People, that were formerly, as will easily appear to any one, who will compare the Accounts we have of it at this time, with Antient History.[121]

Locke's theological argument is at the core of the First Treatise. Note his language of improvement – and especially the connection between improvement and property – when Locke outlines Adam's title by donation. In §39 for example, 'Man's Propriety in the Creatures is nothing but that *Liberty to use them*, which God has permitted, and so Man's property may be altered and enlarged'.[122] Similarly in §40, 'God *gives us all things richly to enjoy*';[123] in §41, 'God ... bid Mankind increase and multiply ... to make use of the Food and Rayment, and other Conveniences of Life, the Materials whereof he had so plentifully provided for them ... to promote the great Design of God, *Increase and Multiply*.'[124]

Locke's imperial language of improvement is profoundly important in his chapter on property. The theological argument about empire forms the basis of Locke's justification of property. He explains that 'God, who hath given the World to Men in common, hath also given them reason to make use of it to the best advantage of Life, and convenience. The Earth, and all that is therein, is given to Men for the Support and Comfort of their being.'[125] Locke's labour theory of property is directed towards making the act of acquiring property part of man's improvement of the earth: 'The *Labour* of his Body, and the *Work* of his Hands, we may say, are properly his. Whatsoever then he removes out of the State that Nature hath provided, and left it in, he hath mixed his *Labour* with, and joined to it something that is his own, and thereby makes it his *Property*.'[126]

David Armitage argues that, when compared with the previous chapters, Locke used a distinctly different language in the chapter on property: 'The language of power and authority, liberty and equality, is strikingly absent from chapter V, whose key terms are instead labour, industry and property. This discontinuity in vocabulary suggests that "Of Property" was composed independently.'[127] While not disputing that Chapter V was composed independently, I think Locke's language is more consistent than Armitage makes out. The language of improvement contains the terms 'labour', 'industry' and 'property', and this language is found not only in Chapter V, but also throughout the First Treatise. In fact, the key Biblical quotation from Paul's Letter to Timothy, that 'God has given us all things richly to enjoy' first quoted in Book 1, §40, is also quoted in Locke's chapter on property: '*God has given us all things richly*, 1 Tim. 6:17 is the Voice of Reason confirmed by Inspiration', Locke writes.[128]

This is not the only instance of continuity between the chapter on property and the theology of the First Treatise. In Chapter V, Locke mentions the very same Biblical pronouncements he did in the First Treatise, about restoring Adam's empire through man's industry and the improvement of the earth. In fact, immediately after quoting Paul's First Letter to Timothy, Locke articulates his theory of property by referring explicitly to the Fall from Eden, and to man's lost empire over the earth which must be regained:

> God, when he gave the World in common to all Mankind, commanded Man also to labour, and the penury of his Condition required it of him. God and his Reason commanded him to subdue the Earth, *i.e* improve it for the benefit of Life, and therein lay out something upon it that was his own, his labour. He that in Obedience to this Command of God, subdued, tilled and sowed any part of it, thereby annexed to it something that was his *Property*.[129]

There could hardly be a more powerful quotation to reveal the fact that man's acquisition of property through labour is inherently part of his restoration of empire over the earth, the very same empire he lost in the Fall from Eden. Throughout the First Treatise the references to God's imperial project for man occur too frequently for each one to be quoted. When explaining the origin of private property, for example, Locke tells us yet again that 'The Law man was under [in the beginning] was rather for *appropriating.* God Commanded, and his Wants forced him to *labour.* That was his *Property* which could not be taken from him.'[130] This is political economy; using the land most effectively for the benefit of England.

I am not suggesting that Locke's consistent use of the imperial language of improvement redates his composition of the chapter. Rather, when understood in the context of the language of improvement, there is more continuity to the *Two Treatises* than historians have recognized. The following statement crystallizes this point. It is taken from the chapter on property but embodies the ideal of empire that shapes the *Two Treatises* as a whole:

> Subduing or cultivating the Earth, and having Dominion, we see are joined together. The one gave Title to the other. So that God, by commanding to subdue, gave Authority so far to appropriate.[131]

This is the very essence of Locke's theory of property. When read in the way I have put forward, the *Two Treatises* become a coherent tract on the restoration of man's *empire* over the earth.

For Locke, political economy was a set of practices enabling England to realize the fruitfulness of the earth. He articulated these practices not only in the *Two Treatises* but throughout his work relating to the colonies. One of the most important aspects of political economy in the Atlantic was trade. In *Some Considerations of the Consequences of the Lowering of Interest, and Raising the Value of Money* (1691), Locke explained how wealth is both gained and lost, and argued that consuming fewer foreign commodities than labour will benefit England. In the language of improvement, Locke outlined his argument in favour of commerce and trade:

> There are but two ways of growing Rich, (*i.e.* of bringing more Riches, and consequently more Plenty of all the conveniences of Life, than what falls to the share of Neighbouring Kingdoms and States) and those two ways of growing Rich, are either Conquest or Commerce ... I think that in our present circumstances, no Body is vain enough to entertain a Thought of our reaping the Profits of the World by Swords ... Commerce therefore is the only way left to us, either for Riches or Subsistence, for this the advantages if our Situation, as well as the Industry and Inclination of our People, bold and skilful at Sea, do naturally fit us.[132]

The text is full of phrases that describe trade as contributing to 'the improvement of the general stock and wealth of the Nation'.[133] It is clear from Locke's manuscripts that he read the Bristol merchant and author John Cary's *Essay on the State of England in Relation to its Trade, its Poor and its Taxes* (1695). Cary argued for the viability of American colonies, and for the protection of English wool textile manufacture, for the confinement of the Irish textile industry to linen, and for the employment of the poor. Locke praised Cary's work as being 'the best discourse I ever read on that subject', the primary reason being that the aim of Cary's proposals was for 'the publick good'. Trade was vital and Cary could not have 'employ[ed] [his] thoughts on a more necessary or usefull subject'.[134] Locke's views on the utility of trade and its role in producing the necessities for the English population were consistent throughout his work. In Locke's essay on trade, composed in 1674, he restated his earlier ideas: 'The chief end of trade is riches and power which beget each other.'[135] Referring once more to the practical arts, Locke mentions the types of employment that are crucial to trade. These include 'husbandry, drapery, mines and navigation'.[136]

Locke's interest in trade and improvement was manifested not only in his interest in the American colonies but also in Ireland. His manuscripts bear witness to this. In Locke MS c30 there is 'An Abstract of the Heads of the Bill for encouraging the Linnen and Hempen Manufacture in Ireland (1697) which was originally enclosed with a letter of William Molyneux to Locke dated 4 October of that year. The tract makes the connection between the charges and prices for the production of linen and hemp and the provision for schools for poor children who would help manufacture the materials. It stipulates an 'account to be made of all Fines &c at the next Count that shall be received and of all necessary Charges & touching the same and the overplus which remaines after the said necessary Charges &c are Dedicated to the Remainder to be Employed in the Educating and Employing Poor Children'.[137] The children who would produce the linen would be provided for by a spinning school in each county 'wherein 40 poor Children are to be fedd, Clothed and Educated in reading repeating without Book the Church Chatechism & to Spinn and Knitt'.[138]

Locke's interest in Ireland as a site for implementing his political economy schemes is evident in a number of tracts that he owned relating to Ireland which are now contained in his manuscript collection. There is, for example, a 'Memorial to the Lords of the Councell of Trade by the Directors of the Royal Lustring Company' (1698) and a 'Copy of a letter of Philip Bayly to George Stead concerning linen manufacture' dated 1697. In Locke MS c30 is a 'Proposall for a Poll Tax of Hemp and Flax in Ireland' dated in 1697, which was endorsed by Locke. In this proposal, observations about Ireland's natural world came together with arguments for how the plantations could be improved. The Proposal begins with some natural observations: 'the soyle of Ireland is found by experience proper for the productions of Hemp and Flax'.[139] It is then calculated that 'there are in Ireland computed to be about 1200 000 people'.[140] The Proposal then assures its readers that 'this will encourage the natives to the establishing the linen manufacture already began, and where some tolerably good is made; and may be improved, not only for their own use, but also to supply England in many kinds as good and usefull as what is ... from Flanders and other parts'.[141]

Locke's papers relating to Virginia are concerned above all with the political economy of improvement. 'Queries to be put to Coll. Henry Hartwell or any other discreet person that knows the constitution of Virginia', dated 1697, asks questions about the nature of the land, and what 'improvements' planters are 'obliged to meet upon their Land'.[142] There is also, in Locke MS e9, a list of 'Some of the Chief Grievances of the present constitution of Virginia with an Essay towards the Remedies thereof'. This tract is about the potential improvements to be made to the colony. The author laments that the country is 'ill peopled' and that 'many other usefull Improvements for yr kingdom of England are neglected, whether of Country were well peopled might be made among the English' and recommends 'the Manufacture of Iron & all other minerals ... the Manufacture of Silks ... the Manufacture of Pot-ash for Soap' and 'the Manufacture of Wheat, Rye, Indian Corn, & all sorts of grain, very usefull for ye supply of our other English plantations'.[143]

Schemes for the education and employment of children were central to improving the world and the advancement of man because they reduced idleness. In his *Some Thoughts Concerning Education*, we find the basis of what Locke was to apply to America. Locke advised: 'I would have him *learn a Trade, a Manual Trade*; nay two or three, but one more particularly.'[144] In this respect, Locke advised '*Gardening* or *Husbandry* in general'.[145] The 'great Men among the Ancients' understood very well how to reconcile manual Labour with Affairs of State ... They were great Captains and Statesmen as well as Husdbandmen.'[146] He speaks of '*Delving, Planting*' as 'profitable Employments'.[147] In a letter to Edward Clarke in 1688, advising him upon how to educate his son, Locke writes: 'I would have your son learn a trade, a handicrafts trade ... gardening, or working in wood, as a carpenter, joiner, or turner ... by being skilled and exercised in the one of them, he will be able to govern or teach his gardener; and by the other contrive, and make a great many curious things of both delight and use.'[148]

This tract on education reveals that Locke used the same language of improvement whether his subject matter was the colonies or personal improvement at home. His vocabulary embodied the idea of empire, which should be restored over external colonies and emulated by a mastery over one's self. The fact that Locke's language of empire remained the same regardless of whether Locke referred to Carolina, Ireland or educating young gentlemen reveals that his language was generic; that is, it embodied the idea of creating dominion wherever it was applied. James Tully has argued that Locke's educational work proposed a new mode of governing the self through the reform of thought and behaviour.[149] We can extend this idea. For Locke, self-improvement was part of the same process of gaining dominion over colonized lands; it was a theory of man's empire. Dominion over land was an extension of having mastery over the self.

Locke's schemes of political economy related to England as well as the colonies. This illustrates my point: it is not that there is a colonial 'dimension' to Locke's work which consists in his writings explicitly relating to colonies, but rather that the vast majority of Locke's work is fundamentally about empire; that is, empire as dominion over nature. When Locke writes about colonies or property we should see that as but one example of a corpus of work that is compellingly imperial. Man's dominion over the world is the abiding intellectual coherence of Locke's work.

In his 'Essay on the Poor Law', composed in 1697 when he was Commissioner on the Board of Trade, Locke proposed replacing individual parishes' authority with that of groups of parishes in order to establish factories in which destitute people would be employed to spin wool. The Law was designed to utilize the population as efficiently as possible, to maximize employment. Note the language of improvement when Locke requests His Majesty 'to require us particularly to consider of some proper methods for setting on work and employing the poor of this kingdom, and making them useful to the public'.[150] He then refers to 'virtue and industry being as constant companions on the one side as vice and idleness are on the other'.[151]

Similarly, in Locke's essay which has been given the title 'For a General Naturalisation', he argued in favour of naturalization on the basis that it would be 'the shortest and easiest way of increasing your people ... [which are] the strength of any country or government'.[152] Furthermore, in a reference to the world's lost Edenic state, Locke made it clear that 'the riches of the world do not lie now as formerly in having large tracts of good land which supplied abundantly the native conveniences of eating and drinking [such as plenty of corn and large flocks and herds. But in trade, which brings in money and with that all things'.[153] A general naturalization of groups of people wishing to settle in England would secure the 'advantage' and 'profit of all their labour'.[154]

Labour was central to Locke's schemes of improvement. In his essay 'Labour', written in 1693, Locke emphasized the personal improvement to be gained from efficient labour in the loss of idleness. Once more we see the theological argument that God has commanded man to labour since the Fall: 'we ought to look on it as a mark of goodness in God that he has put us in this life under a necessity of labour'.[155] This labour is 'a benefit even to the good and the virtuous which are thereby preserved from the ills of idleness'.[156] Labour is necessary to reap the fruits of the earth: 'Half the day employed in useful labour would supply the inhabitants of the earth with the necessaries and conveniences of life', Locke wrote, employing the vocabulary he used so often.[157] Locke made it clear that labour is part of man's project of recovering 'conveniences' from the earth. It is an 'honest and useful industry',[158] 'by which all mankind might be supplied with what the real necessities and conveniency of life demand'.[159] With reference to his ideal of mankind, Locke concluded his essay by stating that 'if the labour of the world were rightly directed and distributed there would be more knowledge, peace, health and plenty in it than now there is. And mankind be much more happy than now it is'.[160]

We have now a new intellectual context for reading the colonial work of John Locke. The idea of improvement, the cornerstone of Locke's philosophy, embodied a theory of empire centred upon the restoration of Adam's dominion over a fruitful, fully populated earth. While Locke scholarship has been advanced immensely by placing his work in a colonial context, the concept of empire deepens our understanding further. The imperial language of improvement is the abiding core of Locke's work; it unites his colonial political economy with his natural philosophy, epistemology and theology.

When Locke died in 1704, England was on the verge of building a maritime empire that would create an increasingly complex set of ideologies. In the century following Locke's death, the Adamic, theological ideal of improving the world died out, while the ideal of improving the world was secularized and became an increasingly important justification for British imperial ventures. Locke was at once the pinnacle of the Adamic idea of empire that this book has charted, yet also perhaps the earliest voice for the imperial ethos of 'improvement'. I suspect that John Locke stood at the crossroads between these two ideologies of the British Empire.

CONCLUSION

A certain historical irony resides in the fact that the concept of empire is a popular subject of historical, political and philosophical discourse, yet the extensive field of British Empire scholarship has insufficiently investigated its subject's conceptual origins. This book attempted to bring the history of early modern natural philosophy to bear upon the intellectual origins of the British Empire. The Biblical ideal of man's plenary empire over nature, central to the work of seventeenth-century natural philosophers, constitutes an intellectual tradition that has been overlooked by the scholarship. The British Empire has a neglected ideological lineage.

For Francis Bacon, writing during the first two decades of the century, there was no necessary connection between man's empire over nature and colonization. By the time John Locke's *Two Treatises of Government* was published in 1689, however, the connection existed. A number of intellectual manoeuvres established a nexus between the idea of man's empire over nature and what became the British Empire of colonies. During the course of the seventeenth century, the agrarian aspect of man's plenary empire – the injunction to cultivate the earth –superseded the epistemological ideal of recovering man's perfect knowledge of nature. It was the agrarian idea of restoring man's dominion over a fully cultivated earth which Locke drew upon in his theory of property.

Ideas of man's original dominion can be conceptualized as an intellectual tradition which, through its own vocabulary, stipulated a coherent theory of empire. The vocabulary of Adam's empire over nature broadens our understanding of the linguistic resources available in the seventeenth century. In doing so, it enables us to make that tradition the subject of further exploration. The fact that the idea of Adam's original empire was fundamental to the work of a number of thinkers from the late sixteenth century to the early eighteenth century – from Bacon to Locke – suggests that we should view 'empire' as a major organizing category of seventeenth-century thought. Indeed, as John Milton's *Paradise Lost* (1667) illustrates, natural philosophical writing was not the only context influenced by the idea of man's original dominion over the earth. Future scholarly research might take the form of a rigorous and interdisciplinary investigation into this tradition of empire in seventeenth-century culture.

A second avenue of further research would build upon the enduring legacy of the tradition of man's empire over nature. Over 150 years after Locke's death in 1704, the Canadian lawyer-historian William Jarvis penned a tract on the rights of the English colonizers to the land in the eastern provinces. Before the arrival of the English, Nova Scotia was 'occupied only by wandering tribes, whose small numbers, roving dispositions and unfitness for duties and employments of civilized life, rendered hopeless the improvement of the soil'.[1] Jarvis was writing in the 1860s, but his language, and the foundation of the claims in his essay, were thoroughly Lockean.[2]

Two decades earlier and 12,000 miles away in the colony of New South Wales, the barrister Richard Windeyer made an analogous claim. The Aboriginal people in the colony, he declared in a public speech in 1842, 'range[d] over' rather than inhabited the land. They 'never tilled the soil, or enclosed it, or cleared any portion of it, or planted a single tree, or grain or root'.[3] The Lockean inheritance of nineteenth-century colonizers and the intellectual history of colonial dispossession is a matter of vigorous contemporary debate among historians, lawyers and philosophers. It is little wonder why. We live in a post-colonial moment in which the dispossession of indigenous people and the origins of nationhood animate popular consciousness. It was with a keen sense of the poignancy of history that, in Australia's most famous legal case for indigenous land rights, the nation's High Court described colonization as leaving a legacy of 'unutterable shame'.[4]

The intellectual lineage of British colonial dispossession is complex. Biblical ideas are entwined with a rich natural law tradition, and both stipulate that cultivating the earth is one of the proper pursuits of mankind.[5] This book identified and charted one of the origins of the British Empire's ideological apparatus in the shape of an ideal of man's plenary empire over nature. In mapping this tradition throughout the seventeenth century, I hope to have provided a starting point for a comprehensive study into the relationship between natural philosophy, natural law and the history of the British Empire, from the seventeenth to the twentieth centuries. This is not just a fruitful project but a necessary one. The present, as much as the past, is haunted by the spectre of empire.

NOTES

Preface

1. S. Muthu, *Enlightenment against Empire* (Princeton, NJ: Princeton University Press, 2003), p. 283.
2. F. Bacon, *The New Organon*, ed. L. Jardine and M. Silverthorne (Cambridge: Cambridge University Press, 2000), Book 1, aphorism CXXIX, p. 101.
3. J. Locke, in P. Laslett (ed.), *Two Treatises of Government* (1960; Cambridge: Cambridge University Press, 1963), I, IV, 40. References to the *Two Treatises* are given as treatise, chapter and section numbers.

Introduction

1. R. Boyle, *Some Considerations touching the Usefulness of Experimental Natural Philosophy*, Part II (1671), in M. H. Hunter and E. B. Davis (eds), *The Works of Robert Boyle*, 14 vols (London: Pickering & Chatto, 2000), vol. 6, p. 406.
2. F. Bacon, *New Atlantis*, in *Francis Bacon: The Major Works*, ed. B. Vickers (1996; Oxford: Oxford University Press, 2002), p. 480.
3. J. Locke, in P. Laslett (ed.), *Two Treatises of Government* (Cambridge: Cambridge University Press, 1963), I, IX, 86. References are given as treatise, chapter and section numbers.
4. R. Yeo, *Encyclopaedic Visions: Scientific Dictionaries and Enlightenment Culture* (Cambridge: Cambridge University Press, 2001), p. 11.
5. Ibid., p. 3.
6. D. Armitage, *The Ideological Origins of the British Empire* (Cambridge: Cambridge University Press, 2000), p. 4.
7. V. Lenin, *Imperialism: The Highest Stage of Capitalism* (1917; Moscow: Progress Publishers, 1982).
8. See for example K. Kautsky, 'Germany, England and the World Policy', trans. E. Crawford, *The Social Democrat*, 4 (1900), pp. 230–6.
9. E. Said, *Culture and Imperialism* (New York: Alfred A. Knopf, 1993).
10. R. Williams, *Keywords: A Vocabulary of Culture and Society* (1976; Oxford: Oxford University Press, 1992), p. 159.
11. Armitage, *Ideological Origins*, p. 3.
12. See for example K. Kupperman, *Providence Island: 1630–1641: The Other Puritan Colony* (Cambridge: Cambridge University Press, 1993); V. D. Anderson, 'New Eng-

land in the Seventeenth Century', in N. Canny (ed.), *The Oxford History of the British Empire (OHBE), Volume 1: The Origins of Empire* (Oxford: Oxford University Press, 1999), pp. 193–217; V. D. Anderson, *New England's Generation: The Great Migration and the Formation of Society and Culture in the Seventeenth Century* (Cambridge: Cambridge University Press, 1991); S. Foster, *The Long Argument: English Puritanism and the Shaping of New England Culture, 1570–1700* (Chapel Hill, NC: University of North Carolina Press, 1991); J. F. Martin, *Profits in the Wilderness: Entrepreneurship and the Founding of New England Towns in the Seventeenth Century* (Chapel Hill, NC: University of North Carolina Press, 1991); T. H. Breen, *Puritans and Adventurers: Change and Persistence in Early America* (New York: Oxford University Press, 1980); N. Salisbury, *Manitou and Providence: Indians, Europeans, and the Making of New England, 1500–1643* (New York: Oxford University Press, 1982); E. S. Morgan, *Visible Saints: The History of a Puritan Idea* (Ithaca, NY: Cornell University Press, 1963).

13. F. A. Yates, *Astraea: The Imperial Theme in the Sixteenth Century* (London: Routledge, 1975), pp. 39, 48.

14. For a comprehensive discussion of the Roman law roots of the concept of empire, see W. Ullmann, 'This Realm of England is an Empire', *Journal of Ecclesiastical History*, 30 (1979), pp. 175–203.

15. Armitage, *Ideological Origins*, p. 40.

16. Ibid., p. 35.

17. Ibid., p. 59.

18. The Scottish attempted to establish the colony in 1695.

19. Armitage, *Ideological Origins*, pp. 7–8.

20. Ibid., p. 94.

21. Ibid., p. 176.

22. 'The Masculine Birth of Time', in B. Farrington, *The Philosophy of Francis Bacon: An Essay on its Development 1603–1609* (Liverpool: Liverpool University Press, 1964), p. 59.

23. This is the case for the Great Bible, the Geneva Bible and also the King James Version, though the latter was not published until 1611 and therefore after Bacon wrote this text. The Septuagint uses the cognate term in Greek *archete*.

24. A. Fitzmaurice, *Humanism and America: An Intellectual History of English Colonisation 1500–1625* (Cambridge: Cambridge University Press, 2003), p. 1.

25. D. B. Quinn, 'Ireland and Sixteenth-Century European Expansion', in T. D. Williams (ed.), *Historical Studies: Papers Read before the Irish Conference of Historians*, 1 (London: Bowes & Bowes, 1958), pp. 20–32.

26. J. H. Elliott, 'Afterword', in D. Armitage and M. J. Braddick (eds), *The British Atlantic World 1500–1800* (Basingstoke: Palgrave Macmillan, 2002), pp. 233–49, on p. 239.

27. This is raised most famously by J. H. Elliott's seminal work *The Old World and the New: 1492–1650* (Cambridge: Cambridge University Press, 1970). This issue, which more recently encompasses the wide literature on the colonial 'Other', is not relevant to the conceptual origins of empire, and thus will not be discussed here.

28. R. Drayton, 'Knowledge and Empire', in P. J. Marshall (ed.), *OHBE, Volume 2: The Eighteenth Century* (Oxford: Oxford University Press, 1998), pp. 231–52.

29. R. Drayton, 'Science, Medicine, and the British Empire', in R. W. Winks (ed.), *The Oxford History of the British Empire, Volume 5: Historiography* (Oxford: Oxford University Press, 1999), pp. 264–76.

30. Ibid., pp. 264–8.

31. J. Chaplin, *Subject Matter: Technology, the Body and Science on the Anglo-American Frontier, 1500–1676* (Cambridge, MA: Harvard University Press, 2001).

32. R. Grove, *Ecology, Climate and Empire: Colonialism and Global Environmental History: 1400–1940* (Cambridge: Cambridge University Press, 1997).

33. R. Grove, *Green Imperialism: Colonial Expansion, Tropical Island Edens and the Origins of Environmentalism, 1600–1800* (Cambridge: Cambridge University Press, 1995).

34. Ibid., p. 3.

35. Ibid., p. 51.

36. R. Drayton, *Nature's Government: Science, Imperial Britain and the Improvement of the World* (2000; New Delhi: Orient Longman, 2005), p. 2.

37. Ibid., p. xvii.

38. D. Mackay, *In the Wake of Cook: Exploration, Science and Empire 1780–1801* (New York: St Martin's Press, 1981); D. P. Miller and H. Reill (eds), *Visions of Empire: Voyages, Botany and Representations of Nature* (Cambridge: Cambridge University Press, 1996); J. Gascoigne, *Science in the Service of Empire: Joseph Banks, the British State and the Uses of Science in the Age of Revolution* (Cambridge: Cambridge University Press, 1998); R. M. MacLeod and M. J. Lewis, *Disease, Medicine and Empire: Perspectives on Western Medicine and the Experience of European Expansion* (London and New York: Routledge, 1988); M. Worboys, 'Science and British Colonial Imperialism, 1895–1940' (unpublished D.Phil. thesis, University of Sussex, 1980).

39. For general studies about the relationship between science and empire, see C. J. Glacken, *Traces on the Rhodian Shore: Nature and Culture in Western Thought From Ancient Times to the End of the Eighteenth Century* (Berkeley, CA: University of California Press, 1967); P. Palladino and M. Worboys, 'Science and Imperialism', *Isis*, 84 (1993), pp. 91–102; N. Reingold and M. Rothenberg (eds), *Scientific Colonialism: A Cross-Cultural Comparison* (Washington, DC: Smithsonian Institution Press, 1987); L. H. Brockway, *Science and Colonial Expansion: The Role of the British Royal Botanical Garden* (New York: Academic Press, 1979); J. M. MacKenzie, *Imperialism and the Natural World* (Manchester: Manchester University Press, 1991); J. M. MacKenzie, *The Empire of Nature: Hunting, Conservation and British Imperialism* (Manchester: Manchester University Press, 1988); S. Harding, *Is Science Multicultural? Postcolonialisms, Feminisns and Epistemologies* (Bloomington, IN: Indiana University Press, 1998), esp. pp. 1–54; P. Petitjean, C. Jami and A. M. Moulin (eds), *Science and Empires: Historical Studies about Scientific Development and European Expansion* (Dordrecht: Kluwer Academic Publishers, 1991); M. A. Alam, 'Science and Imperialism: What is Science?', *Race & Class*, 19 (1978),pp. 241–51; Eric Hobsbawm, *Industry and Empire: From 1750 to the Present Day, Volume 3: The Pelican Economic History of Britain* (1968; Baltimore, MD: Pelican, 1969). Lynn White Jr has made the claim that the Judaeo-Christian preoccupation with man's dominion over nature is the major intellectual source of twentieth-century environmental destruction. See L. White Jr, 'The Historical Roots of Our Environmental Crisis', *Science*, 155 (1967), pp. 1202–7.

40. M. Hunter, *Science and Society in Restoration England* (Cambridge: Cambridge University Press, 1981); Chaplin, *Subject Matter*.

41. A. Cunningham, 'How the *Principia* Got Its Name', *History of Science*, 29 (1991), pp. 377–92, on p. 388.

42. R. K. Merton, *Science, Technology and Society in Seventeenth-Century England* (1938; New York: Fertig, 1970).

43. C. Webster, *The Great Instauration: Science, Medicine and Reform, 1626–1660*, 2nd edn (1975; Bern: Peter Lang, 2002), p. xx.

44. *The Instauratio Magna, Part II: Novum Organon and Associated Texts*, in *The Oxford Francis Bacon, Volume 11*, ed. G. Rees and M. Wakely (Oxford: Clarendon Press, 2004), I:93, p. 151. References to the *New Organon* are given book: aphorism, page.

45. Webster, *The Great Instauration*, p. xx.

46. On the Merton thesis and its critics, see L. Mulligan, 'Civil War Politics, Religion and the Royal Society', in C. Webster (ed.), *The Intellectual Revolution of the Seventeenth Century* (London and Boston, MA: Routledge & Kegan Paul, 1974), pp. 317–39; L. Mulligan, 'Puritans and English Science: A Critique of Webster', *Isis*, 71 (1980), pp. 457–69; I. B. Cohen (ed.), *Puritanism and the Rise of Modern Science: The Merton Thesis* (New Brunswick and London: Rutgers University Press, 1990); G. A. Abraham, 'Misunderstanding the Merton Thesis: A Boundary Dispute between History and Sociology', *Isis*, 74 (1983), pp. 368–87; L. Feuer, 'Science and the Ethic of Protestant Asceticism: A Reply to Professor Robert K. Merton', *Research in Sociology of Knowledge, Sciences and Art*, 2 (1979), pp. 1-23.

47. J. R. Jacob and M. Jacob, 'The Anglican Origins of Modern Science: The Metaphysical Foundations of the Whig Constitution', *Isis*, 71 (1980), pp. 25–67.

48. Ibid., p. 251.

49. J. R. Jacob, 'The Political Economy of Science in Seventeenth-Century England', *Social Research*, 59 (1992), p. 507.

50. Hunter, *Science and Society*.

51. S. Shapin and S. Schaffer, *Leviathan and the Air Pump: Hobbes, Boyle and the Experimental Life* (Princeton, NJ: Princeton University Press, 1985), pp. 3–21. See also S. Shapin, *A Social History of Truth: Civility and Science in Seventeenth-Century England* (Chicago, IL: University of Chicago Press, 1994).

52. Shapin and Schaffer, *Leviathan and the Air Pump*, 21.

53. B. Shapiro, 'Latitudinarianism and Science in Seventeenth-Century England', *Past and Present*, 40 (1968), pp. 16–41; M. Hunter, 'Science and Heterodoxy: An Early Modern Problem Reconsidered', in D. C. Lindberg and R.S Westman (eds), *Reappraisals of the Scientific Revolution* (Cambridge: Cambridge University Press, 1990), pp. 437–60; J. Henry, 'The Scientific Revolution in England', in R. Porter and M. Teich (eds), *The Scientific Revolution in National Context* (Cambridge: Cambridge University Press, 1992), pp. 178–210.

54. See for example M. Poovey, *A History of the Modern Fact: Problems of Knowledge in the Sciences of Wealth and Society* (Chicago, IL: University of Chicago Press, 1998); Shapin and Schaffer, *Leviathan and the Air Pump*; R. W. Serjeantson, 'Proof and Persuasion', in K. Park and L. Daston (eds), *The Cambridge History of Science, Volume 3: Early Modern Science* (Cambridge: Cambridge University Press, 2006), pp. 132–75; J. V. Golinski, 'Language, Discourse, and Science', in R. C Olby, G. N. Cantor, J. R. R. Christie and M. J. S. Hodge (eds), *Companion to the History of Modern Science* (London: Routledge, 1990), pp. 110–23; P. Dear, *Discipline and Experience: The Mathematical Way in the Scientific Revolution* (Chicago, IL: University of Chicago Press, 1995).

55. D. Carey, 'Compiling Nature's History: Travellers and Travel Narratives in the Early Royal Society', *Annals of Science*, 54 (1997), pp. 269–92. Carey and Iliffe look chiefly at travel narratives collected in the New World. On the Royal Society's communications within Europe, see D. S. Lux and H. J. Cook, 'Closed Circles or Open Networks? Com-

municating at a Distance during the Scientific Revolution', *History of Science*, 36 (1998), pp. 179–211.

56. R. Iliffe, 'Foreign Bodies: Travel, Empire and the Early Royal Society of London, Part II: The Land of Experimental Knowledge', *Canadian Journal of History*, 34 (1999), p. 23.

57. Ibid., p. 23.

58. J. A. Bennett and S. Mandlebrote, *The Garden, the Ark, the Tower, the Temple: Biblical Metaphors of Knowledge in Early Modern Europe* (Oxford: Bodleian Library, 1998). On medieval and early modern history of the Genesis quotation 1:28, see J. Cohen, *Be Fertile and Increase, Fill the Earth and Master It: The Ancient and Medieval Career of a Biblical Text* (Cornell: Cornell University Press, 1989); A. Williams, *The Common Expositor: An Account of the Commentaries on Genesis 1527–1633* (Chapel Hill, NC: University of North Carolina Press, 1948).

59. Bennett and Mandlebrote, *The Garden, the Ark, the Tower, the Temple*, p. 8.

60. P. Harrison, '"Fill the Earth and Subdue it": Biblical Warrants for Colonization in Seventeenth-Century England', *Journal of Religious History*, 29 (2005), pp. 3–24.

61. Ibid., p. 6.

62. Quoted in ibid., p. 10.

63. Quoted in ibid., p. 13.

64. Ibid., p. 18.

65. P. Harrison, *The Bible, Protestantism and the Rise of Natural Science* (1998; Cambridge: Cambridge University Press, 2001), p. 4.

66. Ibid., p. 208.

67. P. Findlen, 'Courting Nature', in N. Jardine, J. A. Secord and E. Spary (eds), *Cultures of Natural History* (Cambridge: Cambridge University Press, 1996), pp. 57–74, on p. 58.

68. M. Foucault, *The Order of Things: An Archaeology of the Human Sciences* (New York: Pantheon, 1970).

69. Drayton, 'Knowledge and Empire', p. 245.

70. On the influence of Bacon, see A. Perez-Ramos, 'Bacon's Legacy', in M. Peltonen (ed.), *The Cambridge Companion to Bacon* (Cambridge: Cambridge University Press, 1996), pp. 311–34.

71. F. Bacon, 'Valerius Terminus', in J. Spedding, R. L. Ellis and D. D. Heath, *The Works of Francis Bacon*, 14 vols (London: Longman, 1857–74), vol. 3, p. 222.

72. On the Invisible College, see Webster, *The Great Instauration*, pp. 56–7; C. Webster, 'New Light on the Invisible College: The Social Relations of English Science in the mid-Seventeenth Century', *Transactions of the Royal Historical Society*, 24 (1974), pp. 19–42.

73. On scientific societies during and after the Scientific Revolution, see Webster, *The Great Instauration*; Hunter, *Science and Society*. It is debatable whether the societies emerged as a consequence of the experimental nature of natural philosophy. As J. E. McClellan has shown, informal societies of research, across a number of disciplines, existed in the Renaissance when the major focus of universities was teaching. See J. E. McClellan III, *Science Reorganized: Scientific Societies in the Eighteenth Century* (New York: Columbia University Press, 1985).

74. The first meeting was held on 28 November 1660. Charles II, while serving as the figurehead, provided no financial support.

75. On the establishment of the Royal Society and its teething problems, see Hunter, *Science and Society*, and Hunter, *Establishing the New Science: The Experience of the Early Royal Society* (Woodbridge: Boydell & Brewer, 1989), especially ch. 5, pp. 156–84.

76. Hunter, *Establishing the New Science*, ch, 3, pp. 73–121.

77. Ibid., p. 93, n. 14.

78. Ibid., p. 139.

79. J. Locke, in P. H. Nidditch (ed.), *Essay Concerning Human Understanding* (Oxford: Clarendon Press, 1975), Book 4, XII, 10.

80. J. Evelyn, *Elysium Britannicum*, BL, Add. MSS 78342–78344. The work was published posthumously.

81. John Evelyn quoted in J. Prest, *The Garden of Eden: The Botanic Garden and the Re-Creation of Paradise* (New Haven, CT: Yale University Press, 1981), p. 47.

82. Ibid., p. 1; also on gardens, see A. Cunningham, 'The Culture of Gardens', in Jardine et al. (eds), *Cultures of Natural History*, pp. 38–56, esp. pp. 41–9.

83. Bennett and Mandlebrote, *The Garden, the Ark, the Tower, the Temple*, p. 52.

84. P. Seed, *Ceremonies of Possession: Europe's Conquest of the New World 1492–1640* (Cambridge: Cambridge University Press, 1995).

85. John Cotton, quoted in ibid., p. 30.

86. R. Hakluyt, 'Discourse of Western Planting', in E. G. R. Taylor, *The Original Writings and Correspondence of the two Richard Hakluyts*, 2 vols (London: Hakluyt Society, 1935).

87. *Mabo and Others* v. *Queensland* (no. 2) (1992), 175, *Commonwealth Law Reports* 1; F.C. 92/014. The doctrine of property through which Aboriginal people were dispossessed is often termed *terra nullius* in the Australian context. It has been shown recently, however, that the term *terra nullius* was not used during early settlement. Despite this, the intellectual tradition of natural law (and, I will suggest, the Adamic idea of empire) upon which *terra nullius* was based, was the intellectual framework through which the British understood and justified their actions. See H. Reynolds, *The Law of the Land*, 2nd edn (Melbourne: Penguin, 1992).

88. See K. O. Kupperman, *Roanoke: the Abandoned Colony* (Totowa: Rowman & Allanheld, 1984).

89. On the moral philosophy of Jacobean colonization, see Fitzmaurice, *Humanism and America*, ch. 3, pp. 58–101.

90. Fitzmaurice, *Humanism and America*, p. 59.

91. J. H. Rose, A. P. Newton and E. A. Benians (eds), *The Cambridge History of the British Empire, Volume 1: The Old Empire from the Beginnings to 1783* (Cambridge: Cambridge University Press, 1929), p. 274.

92. On England's Navigation Acts and their enforcement, see T. C. Barrow, *Trade and Empire: The British Customs Service in Colonial America, 1660–1775* (Cambridge, MA: Harvard University Press, 1967); M. J. Braddick, 'The English Government, War, Trade and Settlement 1625–1688', in Canny (ed.), *OHBE, 1: Origins of Empire*, pp. 286–308; G. E. Aylmer, 'Navy, State, Trade, and Empire', in Canny (ed.), *OHBE, II: Origins of Empire*, pp. 467–80.

93. Figure quoted in N. Ferguson, *Empire: How Britain Made the Modern World* (Penguin: London, 2004), p. 63.

94. J. Locke, 'Understanding', in *Locke: Political Essays*, ed. M. Goldie (Cambridge: Cambridge University Press, 1997), pp. 264–5.

1 'In a Pure Soil'

1. F. Bacon, *New Atlantis*, in *Francis Bacon: The Major Works*, ed. Vickers, p. 480.

2. 'The Masculine Birth of Time', in B. Farrington, *The Philosophy of Francis Bacon: An Essay on its Development from 1603–1609* (Liverpool: Liverpool University Press, 1964), p. 59.

3. D. Armitage, *The Ideological Origins of the British Empire* (Cambridge: Cambridge University Press, 2000), p. 8.

4. On the Henrician tradition, see W. Ullmann, 'This Realm of England is an Empire', *Journal of Ecclesiastical History*, 30 (1979), pp. 175–203.

5. A. Fitzmaurice, *Humanism and America: An Intellectual History of English Colonization 1500–1625* (Cambridge: Cambridge University Press, 2003), p. 2. Also on Bacon and the colonies, see C. M. Andrews, *The Colonial Period of American History*, 4 vols (New Haven, CT: Yale University Press, 1934–8), vol. 1, pp. 72, 85, 303–4.

6. Fitzmaurice, *Humanism and America*, pp. 166, 169, 187.

7. Armitage, *Ideological Origins*, p. 57.

8. Ibid., p. 59.

9. Webster, *The Great Instauration*, p. 12.

10. Ibid., p. 18.

11. Ibid., p. 24.

12. Ibid., p. 45.

13. Farrington, *Philosophy of Francis Bacon*, pp. 21–7.

14. Ibid., p. 28.

15. A. Perez-Ramos, *Francis Bacon's Idea of Science and the Maker's Knowledge Tradition* (Oxford: Clarendon Press, 1988), pp. 84–5.

16. B. Vickers, 'Francis Bacon and the Progress of Knowledge', *Journal of the History of Ideas*, 53 (1992), p. 514.

17. A. Perez-Ramos, 'Bacon's Forms and the Maker's Knowledge Tradition', in M. Peltonen (ed.), *The Cambridge Companion to Bacon* (Cambridge: Cambridge University Press, 1996), p. 115.

18. C. Whitney, 'Francis Bacon's *Instauratio:* Dominion of and over Humanity', *Journal of the History of Ideas*, 50 (1989), p. 371.

19. Ibid., p. 371.

20. Ibid., pp. 373–4.

21. Ibid., p. 377.

22. Ibid., p. 381.

23. P. Rossi, *Francis Bacon: From Magic to Science,* trans. S. Rabinovitch (London: Routledge, 1968), p. 26. See also P. Rossi, 'Bacon's Idea of Science', in Peltonen (ed.), *Cambridge Companion to Bacon*, pp. 25–46.

24. Rossi, *Francis Bacon: From Magic to Science*, p. 19.

25. Ibid., p. 19.

26. S. Gaukroger, *Francis Bacon and the Transformation of Early Modern Philosophy* (Cambridge: Cambridge University Press, 2001).

27. Ibid., p. 5.

28. Ibid., p. 5.

29. Ibid., pp. 6–8.

30. J. Martin, *Francis Bacon, the State and the Reform of Natural Philosophy* (Cambridge: Cambridge University Press, 1992), p. 106.

31. Gaukroger, *Francis Bacon and the Transformation of Early Modern Philosophy*, p. 4.

32. Ibid., p. 9.

33. M. Peltonen, 'Politics and Science: Francis Bacon and the True Greatness of States', *Historical Journal*, 35 (1992), pp. 279–305.

34. H. B. White, 'Bacon's Imperialism', *American Political Science Review*, 102 (1958), pp. 481, 489; H. B. White, *Peace Among the Willows: The Political Philosophy of Francis Bacon* (The Hague: Martinus Nijhoff, 1968).

35. In this tradition, see Martin, *Francis Bacon*; J. Weinberger, *Science, Faith and Politics: Francis Bacon and the Utopian Roots of Modern Age, a Commentary on Bacon's Advancement of Learning* (Ithaca, NY: Cornell University Press, 1985); C. Whitney, *Francis Bacon and Modernity* (New Haven, CT: Yale University Press, 1986); C. Hill, *Intellectual Origins of the English Revolution* (Oxford: Clarendon University Press, 1965).

36. P. Findlen, 'Francis Bacon and the Reform of Natural History in the Seventeenth Century', in D. R. Kelley (ed.), *History and the Disciplines: The Reclassification of Knowledge in Early Modern Europe* (Rochester, NY: University of Rochester Press, 1997), pp. 239–60, on p. 244.

37. Cited in Perez-Ramos, *Francis Bacon's Idea of Science*, p. 106.

38. Ibid., p. 106.

39. F. Bacon, *Cogitata et Visa*, in J. Spedding, R. L. Ellis and D. D. Heath, *The Works of Francis Bacon*, 14 vols (London: Longman, 1857–74), vol. 3, p. 612.

40. Findlen, 'Francis Bacon and the Reform of Natural History', p. 248.

41. The Second Charter is reprinted in W. Hening, *Statutes at Large, Being a Collection of all the Laws of Virginia, from the First Session of the Legislature in the Year 1619*, 13 vols (1819–23; Charlottesville, VA: University Press of Virginia, 1969), vol. 1, pp. 80–98.

42. L. Jardine and M. Silverthorne, 'Introduction', in Bacon, *The New Organon*, ed. Jardine and Silverthorne, p. xiii.

43. F. Bacon, Preliminary to the *Instauratio Magna*, in *The Instauratio Magna, Part II: Novum Organon and Associated Texts*, in *The Oxford Francis Bacon, Volume 11*, ed. G. Rees and M. Wakely (Oxford: Clarendon Press, 2004), p. 45.

44. F. Bacon, *New Organon*, in *The Oxford Francis Bacon, Volume 11*, ed. Rees and Wakely, I:84, 133. References to the *New Organon* are given book: aphorism, page.

45. On the significance of the symbolism of Solomon's Temple, see J. A. Bennett and S. Mandlebrote, *The Garden, the Ark, the Tower, the Temple: Biblical Metaphors of Knowledge in Early Modern Europe* (Oxford: Bodleian Library, 1998).

46. D. H. Sacks, 'Rebuilding Solomon's Temple: Richard Hakluyt and Empire in the "Age of Discovery"', unpublished conference paper, 74th Anglo-American Conference on States and Empires, 8 July 2005 (cited with permission).

47. Ibid.

48. Bacon, *New Organon*, in *The Oxford Francis Bacon, Volume 11*, ed. Rees and Wakely, I:3, 65.

49. For example, in ibid., II:4, 203, where Bacon states, 'the routes to human power and knowledge lie very close together and are almost identical'.

50. Ibid., II:52, 447.

51. P. Harrison, 'Curiosity, Forbidden Knowledge, and the Reformation of Natural Philosophy in Early Modern England', *Isis*, 92 (2001), p. 266.

52. Bacon, Preliminary to the *Instauratio Magna*, in *The Oxford Francis Bacon, Volume 11*, ed. Rees and Wakely, p. 23.

53. F. Bacon, 'Filium Labyrinth', in J. Spedding, R. L. Ellis and D. D. Heath (eds), with a new introduction by Graham Rees, *Collected Works of Francis Bacon*, 7 vols (London: Routledge/Thoemmes, 1996), vol 3, part 2, p. 501. Although a facsimile copy of the

original Spedding et al. edition, the volume and page numbering differs. To prevent confusion between the two Spedding editions, I will refer to the new facsimile version henceforth as Rees, Spedding et al.

54. F. Bacon, *The Advancement of Learning*, in *The Oxford Francis Bacon, Volume 4*, ed. M. Kiernan (Oxford: Clarendon Press, 2000), p. 36.

55. Bacon, 'The Masculine Birth of Time', in Farrington, *The Philosophy of Francis Bacon*, p. 59 (my emphasis).

56. F. Bacon, 'Temporis Partus Masculus', in *Collected Works of Francis Bacon*, ed. Rees, Spedding et al., vol 3, part 2, p. 527.

57. Bacon, 'The Masculine Birth of Time', in Farrington, *The Philosophy of Francis Bacon*, p. 62 (my emphasis).

58. Bacon, 'Temporis Partus Masculus', in *Collected Works of Francis Bacon*, ed. Rees, Spedding et al., vol 3, part 2, p. 528 (my emphasis).

59. Bacon, 'The Masculine Birth of Time', in Farrington, *The Philosophy of Francis Bacon*, p. 70.

60. Bacon, 'Temporis Partus Masculus', in *Collected Works of Francis Bacon*, ed. Rees, Spedding et al., vol 3, part 2, p. 537 (original emphasis).

61. F. Bacon, 'The Wisdom of the Ancients', in *Works*, ed. Spedding et al., vol. 6, p. 696.

62. Ibid., p. 699.

63. Ibid., p. 756.

64. Ibid., p. 757.

65. Ibid., p. 679.

66. Ibid., p. 757.

67. Ibid., p. 679.

68. Ibid., p. 758.

69. Ibid., p. 680.

70. Ullmann, 'This Realm of England is an Empire', p. 176.

71. 'Certain Articles or Considerations, touching the union, of the Kingdomes of England, and Scotland', in W. Rawley, *Resuscitatio, or Bringing into Publick Light Severall Pieces of the Works, Civil, Historical, Philosophical, and Theological, Hitherto Sleeping, of the Right Honourable Francis Bacon* (London, 1657), p. 212.

72. F. Bacon, *De Augmentis Scientiarum*, in *Works*, ed. Spedding et al., vol. 5, p. 31.

73. F. Bacon, 'Of Empire', in *The Essayes or Counsels, Civill and Moral*, in *Oxford Francis Bacon, Volume 15*, ed. M. Kiernan (Oxford: Clarendon Press, 2000), pp. 58–63.

74. Bacon, *New Organon*, in *Oxford Francis Bacon, Volume 11*, ed. Rees and Wakely, I:84, 69.

75. Ibid., I:93, 151.

76. Ibid., I: 72, 117.

77. F. Bacon, 'In Praise of Knowledge', in *Works*, ed. Spedding et al., vol. 8, pp. 125–6.

78. Bacon, *New Organon*, in *Oxford Francis Bacon, Volume 11*, ed. Rees and Wakely, I:129, 195.

79. Ibid., II: 36, 321.

80. F. Bacon, 'Certain Considerations Touching the Plantation in Ireland', in *Works*, ed. Spedding et al., vol. 11, p. 115.

81. Cicero, *De Officiis*, ed. M. T. Griffin and E. M. Atkins (Cambridge: Cambridge University Press, 1996), p. 19.

82. Sallust, *War with Catiline*, trans. P. McGushin (Bristol: Bristol Classical Press, 1980), p. 11.

83. Ibid., p. 21.
84. 'A letter of advice written by Sir Francis Bacon to the Duke of Buckingham', in *Works*, ed. Spedding et al., vol. 8, p. 21.
85. Ibid., p. 21.
86. 'Of Plantations', in *Essayes*, ed. Kiernan, p. 106.
87. 'Considerations Touching the Queen's Service in Ireland', in *Works*, ed. Spedding et al., vol. 10, p. 47. Note that in the seventeenth century, 'reduction' was a term for 'reform'.
88. F. Bacon, 'Letter to Cecil on the State of Ireland, inciting him to embrace the care of reducing that kingdom to civility', in ibid., vol. 10, p. 45.
89. Bacon, 'Advice to Sir George Villiers', in ibid., vol. 13, p. 21.
90. 'Of Plantations', in Kiernan (ed.), *Essayes*, p. 108.
91. Bacon, 'Considerations Touching the Queen's Service in Ireland', in *Works*, ed. Spedding et al., vol. 10, p. 49.
92. Ibid.
93. Ibid., p. 51.
94. F. Bacon, 'Speech for General Naturalisation', in ibid., vol. 10, p. 315.
95. See, for example, J. H. M Salmon, 'Seneca and Tacitus in Jacobean England', in L. L. Peck (ed), *The Mental World of the Jacobean Court* (Cambridge: Cambridge University Press, 1991); P. Burke, 'Tacitism', in T. A. Drey (ed.), *Tacitus* (London: Routledge, 1969), pp. 149–71; M. Peltonen, *Classical Humanism and Republicanism in English Political Thought 1570–1640* (Cambridge: Cambridge University Press, 1995); Fitzmaurice, *Humanism and America*.
96. Bacon, 'Speech for General Naturalisation', in *Works*, ed. Spedding et al., vol. 10, p. 315.
97. F. Bacon, 'Apology Concerning Essex', in ibid., vol. 10, p. 146.
98. Bacon, 'Advice to Sir George Villiers', in ibid., vol. 13, p. 10.
99. F. Bacon, 'Of the True Greatness of the Kingdom of Britain', in ibid., vol. 7, p. 56.
100. Ibid., p. 56.
101. Ibid., p. 56.
102. F. Bacon, *Sylva Sylvarum*, in *Works*, ed. Spedding et al., vol. 2, pp. 472–3.
103. Ibid., p. 518.
104. F. Bacon, *Historia Ventorum*, in *Works*, ed. Spedding et al., vol. 5, p. 147.
105. Ibid., p. 158.
106. Ibid., p. 161.
107. Ibid., p. 167.
108. Ibid. pp. 168–9.
109. Ibid., p. 172.
110. Ibid., p. 152.
111. 'Translation of the Historia Vitae et Mortis', in *Works*, ed. Spedding et al., vol. 5, p. 272.
112. Ibid., p. 284.
113. Ibid., p. 296.
114. 'De Fluxu et Refluxu Maris', in *Philosophical Studies c. 1611–1619*, in *Oxford Francis Bacon, Volume 6*, ed. G. Rees (Oxford: Clarendon Press, 1996), p. 81.
115. 'Preparations Towards a Natural and Experimental History', in *The Oxford Francis Bacon, Volume 11*, ed. Rees and Wakely, p. 451.
116. 'De Fluxu et Refluxu Maris', in *Philosophical Studies*, ed. Rees, p. 89.
117. D. Albanese, 'New Atlantis and the Uses of Utopia', *English Literary History*, 57 (1990), p. 504.

118. A. Boesky, *Founding Fictions: Utopias in Early Modern England* (Athens: University of Georgia Press, 1996), pp. 56–84; A. Boesky, 'Bacon's *New Atlantis* and the Laboratory of Prose', in E. Fowler and R. Greene (eds), *The Project of Prose in Early Modern Europe and the New World* (Cambridge: Cambridge University Press, 1997), pp. 138–53.

119. C. Whitney, 'Merchants of Light: Science as Colonization in the *New Atlantis*', reprinted in W. Sessions (ed.), *Francis Bacon's Legacy of Texts* (New York: AMS Press, 1990), pp. 255–68.

120. C. Jowitt, 'Colonialism, Jewishness and Politics in Bacon's *New Atlantis*', in B. Price (ed.), *Francis Bacon's New Atlantis: New Interdisciplinary Essays* (Manchester: Manchester University Press, 2002), p. 130.

121. R. Serjeantson, 'Natural Knowledge in the *New Atlantis*', in ibid., p. 97.

122. On the utopian tradition, see J. C. Davis, *Utopia and the Ideal Society: A Study of English Utopian Writing 1516–1700* (Cambridge: Cambridge University Press, 1981).

123. Bacon, *New Atlantis*, in *Francis Bacon: The Major Works*, ed. Vickers, pp. 471–2.

124. Ibid., pp. 471–2.

125. Ibid., p. 471.

126. Ibid., p. 480.

127. Serjeantson, 'Natural Knowledge', p. 85.

128. Bacon, *New Atlantis*, in *Francis Bacon: The Major Works*, ed. Vickers, pp. 483–4.

129. Ibid., p. 487.

130. M. Poovey, *A History of the Modern Fact: Problems of Knowledge in the Sciences of Wealth and Society* (Chicago, IL: University of Chicago Press, 1998), esp. pp. 96–103; B. Shapiro, *A Culture of Fact: England 1550–1720* (Ithaca, NY: Cornell University Press, 2000).

2 Restoring Eden in America

1. The Irish Company (1611); East India Company (1614); Somers Islands Company (1615); and New River Company (1619). See L. R. Muir and J. A. White (eds), *Materials for the Life of Nicholas Ferrar: A Reconstruction of John Ferrar's Account on his Brother's Life Based on All the Surviving Copies* (Leeds: Leeds Philosophical and Literary Society, 1996). Nicholas was John's brother; see A. L. Maycock, *Chronicles of Little Gidding* (London: SPCK, 1954).

2. Recent work has placed the Hartlib Circle in a millenarian and European context. See J. E. Force and R. H. Popkin (eds) *Millenarianism and Messianism in Early Modern European Culture, Volume 3: The Millenarian Turn: The Millenarian Contexts of Science, Politics and Everyday Anglo-American Life in the Seventeenth and Eighteenth Centuries* (Dordrecht: Kluwer Academic Publishers, 2001); J. T. Young, *Faith, Medical Alchemy and Natural Philosophy: Johann Moriaen, Reformed Intelligencer, and the Hartlib Circle* (Aldershot: Ashgate, 1998). See also the earlier work on the Hartlib Circle, including Hugh Trevor-Roper who placed the members in the context of the reformation religious turmoil of the 1620s: H. Trevor-Roper, *Religion, the Reformation and Social Change, and Other Essays*, 2nd edn (London: Macmillan, 1967); G. H Turnbull, *Samuel Hartlib: A Sketch of His Life and His Relations to J. A. Comenius* (Oxford: Oxford University Press, 1920).

3. Turnbull, *Samuel Hartlib*, p. 14.

4. In treating Hartlib's colleagues as a 'circle' I am following the lead of Mark Greengrass, Michael Leslie and Timothy Raylor and their recent edited collection: M. Greengrass, M. Leslie and T. Raylor (eds), *Samuel Hartlib and Universal Reformation: Studies in Intellectual Communication* (Cambridge: Cambridge University Press, 1994).

5. J. Peacey, 'Seasonable Treatises: A Godly Project of the 1630s', *English Historical Review*, 113:452 (June 1998), pp. 667–9, on p. 667.

6. Ibid., p. 668. See K. O. Kupperman, *Providence Island 1630–1641: The Other Puritan Colony* (Cambridge: Cambridge University Press, 1993).

7. A number of recent studies also characterize the Hartlib Circle's work as natural philosophy. Jim Bennett and Scott Mandlebrote point out in the introduction to their study that the issues discussed by the Hartlib Circle were various and included 'educational theory, divinity or natural philosophy': see J. A. Bennett and S. Mandlebrote, *The Garden, the Ark, the Tower, the Temple: Biblical Metaphors of Knowledge in Early Modern Europe* (Oxford: Bodleian Library, 1998), p. 157. There is a subsection in Greengrass, Leslie and Raylor, *Samuel Hartlib and Universal Reformation*, entitled 'Natural Philosophies'. A. Clericuzio, 'New Light on Benjamin Worsley's Natural Philosophy' in the same volume deals explicitly with 'Benjamin Worsley's natural philosophy'.

8. S. Hartlib, *The Compleat Husband-Man: or, A Discourse of the Whole Art of Husbandry both Forraign and Domestick* (London, 1659), pp. 81–2.

9. H. McD. Beckles, '"The Hub of Empire": The Caribbean and Britain in the Seventeenth-century', in N. Canny (ed.), *The Oxford History of the British Empire (OHBE), Volume 1: The Origins of Empire* (Oxford: Oxford University Press, 1999), pp. 218–40.

10. D. H. Sacks, *The Widening Gate: Bristol and the Atlantic Economy, 1450–1700* (Berkeley, CA: University of California Press, 1991); R. Brenner, *Merchants and Revolution: Commercial Change, Political Conflict and London's Overseas Traders, 1550–1650* (Princeton, NJ: Princeton University Press, 1993).

11. On England's Navigation Acts and their enforcement, see Barrow, *Trade and Empire*; M. J. Braddick, 'Government, War, Trade and Settlement 1625–1688', in Canny (ed), *OHBE, 1: Origins of Empire*, pp. 286–308; G. E. Aylmer, 'Navy, State, Trade, and Empire', in Canny (ed), *OHBE, 1: Origins of Empire*, pp. 467–80.

12. Digges was governor between *c.* 1652 and 1660.

13. On Comenius, see D. Capkova, 'Comenius and His Ideals: Escape from the Labyrinth', in Greengrass et al., *Samuel Hartlib*, pp. 75–91; H. Hotson, 'Philosophical Pedagogy in Reformed Central Europe between Ramus and Comenius: a Survey of the Continental Background of the "Three Foreigners"', in Greengrass et al., *Samuel Hartlib*, pp. 29–50; J. Privratska and V. Privratsky, 'Language as the Product and Mediator of Knowledge: The Concept of J. A. Comenius', in Greengrass et al., *Samuel Hartlib*, pp. 162–73.

14. Document 7/109 quoted in Introduction to the *The Hartlib Papers*, 2nd edn (Sheffield: HROnline, 2002). The *Hartlib Papers* shall henceforth be abbreviated to *HP*; D. Stimson, 'Hartlib, Haak and Oldenburg: Intelligencers', *Isis*, 31 (1940), pp. 309–26.

15. Webster, *The Great Instauration*, p. 44; See also W. H. G. Armytage, 'The Early Utopists and Science in England', *Annals of Science*, 12 (1956), pp. 247–54. For Webster's other work on the Hartlib Circle, see 'The Authorship and Significance of *Macaria*', *Past and Present*, 56 (1972), pp. 34–49; *Samuel Hartlib and the Advancement of Learning* (Cambridge: Cambridge University Press, 1970); and Webster (ed.), *Utopian Planning and the Puritan Revolution: Gabriel Plattes, Samuel Hartlib and Macaria* (Oxford: Wellcome Unit for the History of Medicine, 1979).

16. S. Hartlib, *The Commonwealth of Bees* (London, 1655), p. 44.

17. A. Speed, *Adam out of Eden, or an Abstract of Divers excellent Experiments Touching the Advancement of Husbandry* (London, 1659).

18. *HP*, 46/5/8A.

19. Ibid., 51/50B.

20. S. Hartlib, *An Essay for Advancement of Husbandry-Learning: or Propositions for the Erecting Colledge of Husbandry* (London, 1651).
21. P. Seed, *Ceremonies of Possession in Europe's Conquest of the New World, 1492–1640* (Cambridge: Cambridge University Press, 1995).
22. *HP*, 15/2/62B.
23. W. Bullock, *Virginia Partially Examined, and Left to Public View, to be Considered by all Judicious and Honest Men* (London, 1649).
24. On Bullock the elder's dealings in Virginia, see J. T. Kneebone, J. J. Looney, B. Tarter and S. Gioia Treadway (eds), *Dictionary of Virginia Biography, Volume 2* (Richmond: Library of Virginia, 2001), pp. 391–2.
25. *HP*, 28/1/18A.
26. Ibid.
27. On the Invisible College, see C. Webster, 'Benjamin Worsley: Engineering for Universal Reform from the Invisible College to the Navigation Act', in Greengrass et al., *Samuel Hartlib*, pp. 213–35.
28. On Benjamin Worsley, see G. H. Turnbull, *Hartlib, Dury and Comenius: Gleanings from Hartlib's Papers* (Liverpool: University of Liverpool Press and Hodder & Stoughton, 1947); T. C. Barnard, *Cromwellian Ireland: English Government and Reform in Ireland, 1649–1660* (Oxford: Oxford University Press, 1975); J. C. Sainty (ed.), *Officials of the Board of Trade, 1660–1870* (London: Athlone Press, 1974); Greengrass et al., *Samuel Hartlib*; A. Clericuzio, 'New Light on Benjamin Worsley's Natural Philosophy', pp. 236–46.
29. Benjamin Worsley to John Dury, 27 July 1653, *HP*, 32/2/18A.
30. S. M. Kingsbury (ed.), *The Records of the Virginia Company of London*, 4 vols (Washington, DC: Government Printing Office, 1906–35).
31. See for example John Dury [?] to Benjamin Worsley, 31 August 1649, *HP*, 1/2/14A–15B; Worsley to W. Strickland, n.d., *HP*, 61/8/1A–3B. On the situation in the American colonies, see J. Horn, *Adapting to a New World: English Society in the Seventeenth-Century Chesapeake* (Chapel Hill, NC: University of North Carolina Press, 1994).
32. P. Thompson, 'William Bullock's Strange Adventure: A Plan to Transform Seventeenth-Century Virginia', *William and Mary Quarterly*, 61 (2004), pp. 107–28, on p. 16.
33. Copy extracts, John Beale and Others to Hartlib, May 1658–June 1658, *HP*, 52/78A.
34. G. Plattes and S. Hartlib, *A Description of the Famous Kingdome of Macaria* (London, 1641), reprinted in Webster, *Samuel Hartlib and the Advancement of Learning*, pp. 81–2.
35. Ibid., p. 82.
36. Ibid., p. 83.
37. See 'The Reformed Virginian Silkworm Or, a Rare and New Discovery of A Speedy way, and Easie Means, Found Out by a Young Lady in England, She Having Made Full Proof Thereof in May, 1652, For the Feeding of Silk-worms in the Woods, on the Mulberry-Tree-leaves in Virginia', in Hartlib, *Commonwealth of Bees*, p. 25.
38. Ferrar, in 'The Reformed Virginian Silkworm', in Hartlib, *Commonwealth of Bees*, sig. A1 r–v.
39. Ibid., p. 4.
40. Ibid., pp. 5–6.
41. Ibid., p. 13.
42. John Beale to Samuel Hartlib, 14 December 1658, *HP*, 51/43A–B.

43. B. Worsley, 'Profits Humbly Presented to this Kingdom', in Hand B, undated, *HP*, 15/2/61A–B.
44. Ibid., 15/2/62B.
45. Ibid., 15/2/62B.
46. Greengrass et al., *Samuel Hartlib*, 18.
47. On Comenius's language schemes and their influence, see U. Eco, *The Search for the Perfect Language* (1993; Oxford: Blackwell, 1995).
48. R. F. Young, *Comenius in England* (London: Oxford University Press, 1934).
49. R. F. Young, *Comenius and the Indians in New England* (London: King's College, 1929).
50. On John Eliot, see R. W. Cogley, *John Eliot's Mission to the Indians before King Philip's War* (Cambridge, MA: Harvard University Press, 1999); O. E. Winslow, *John Eliot, 'Apostle to the Indians'* (Boston, MA: Houghton Mifflin, 1968).
51. *HP*, 28/2/1A.
52. John Beale to Samuel Hartlib, 11 March 1659, *HP*, 51/76A.
53. *HP*, 30/4/56A
54. John Beale to Samuel Hartlib, 11 March 1659, *HP*, 51/76A.
55. Ibid., 15/2/66A.
56. Ibid., 29/8/6B.
57. Copy letter in Scribal Hand B, Robert Child to Hartlib, 24 December 1648, *HP*, 39/2/5A.
58. *HP*, 15/2/61A–B.
59. S. Hartlib, *Samuel Hartlib his Legacie or An Enlargement of the Discourse of Husbandry used in Brabant and Flaunders; Wherein are bequeathed to the Common-Wealth of England more Outlandish and Domestick Experiments and Secrets in reference to Universall Husbandry* (London, 1651), p. 24.
60. Ibid., pp. 46–7.
61. Ibid., pp. 63–4.
62. Ibid., p. 69.
63. P. Anstey, 'Hartlib and Starkey Rekindled', *Metascience*, 13 (2004), pp. 112–15; W. R. Newman, *Gehennical Fire; the Lives of George Starkey, an American Alchemist in the Scientific Revolution* (1994; Chicago, IL: University of Chicago Press, 2002).
64. Hartlib, *Commonwealth of Bees*, p. 21.
65. Ibid., p. 22.
66. Ibid., p. 23.
67. Ibid., p. 24.
68. Ibid., p. 24.
69. Ibid., p. 38.
70. Hartlib, *Compleat Husband-Man*, p. 18.
71. Ibid., p. 22.
72. Ibid., p. 36.
73. Ibid., p. 74.
74. Ibid., p. 49.
75. Ibid., p. 52.
76. Ibid., p. 62.
77. Ibid., p. 68.
78. Ibid., pp. 68–9.
79. Ibid., p. 74.

80. Ibid., p. 61.
81. Ibid., p. 61.
82. Ibid., p. 71.
83. Ibid., pp. 78–80.
84. Ibid., p. 62.
85. Ibid., pp. 105–6.
86. S. Hartlib, *Cornucopia: A Miscellanium of Lucriferous and most Fructiferous Experiments, Observations, and Discoveries, Immethodically Distributed; to be Really Demonstrated and Communicated in All Sincerity* (London, 1652).
87. Ibid., p. 1.
88. Ibid., p. 9.
89. Ibid., p. 5.
90. Ibid., p. 8.
91. 'The Reformed Virginian Silkworm', in Hartlib, *Commonwealth of Bees*, sig. A3 r.
92. Ibid., p. 2.
93. Ibid., p. 5.
94. On the *Commonwealth of Bees*, see M. Leslie and T. Raylor (eds), *Culture and Cultivation in Early Modern England* (Leicester: Leicester University Press, 1992).
95. Hartlib, *Commonwealth of Bees*, p. 2. For two excellent studies of the metaphor of the beehive and its importance in the American colonies, see K. O. Kupperman, 'The Beehive as a Model for Colonial Design', in K. O. Kupperman, *America in European Consciousness: 1493–1750* (Chapel Hill, NC: University of North Carolina Press, 1995), pp. 272–92; B. Wilson, *The Hive: The Story of the Honeybee and Us* (London: John Murray, 2004).
96. Hartlib, *Cornucopia*, p. 2.
97. Hartlib, *Commonwealth of Bees*, p. 50.
98. Ibid., p. 51.
99. Ibid., p. 54.
100. Ibid., p. 8.
101. L. Daston, 'The Moral Economy of Science', in A. Thackray (ed.), *Constructing Knowledge in the History of Science* (Chicago, IL: University of Chicago Press, 1995); see also Ian Hacking's discussion of the term in *Historical Ontology* (Cambridge, MA: Harvard University Press, 2002), pp. 7–11.
102. P. Dear, *Discipline and Experience: The Mathematical Way in the Scientific Revolution* (Chicago, IL: University of Chicago Press, 1995). See also Shapiro, *A Culture of Fact*.
103. L. Daston uses the phrase frequently. See for example 'The Moral Economy of Science' *Osiris*, 2nd series, 10 (1995), pp. 2–24.
104. M. Poovey, *A History of the Modern Fact: Problems of Knowledge in the Sciences of Wealth and Society* (Chicago, IL: University of Chicago Press, 1998), p. 9.
105. Hartlib, *An Essay for the Advancement of Husbandrie-Learning*, p. 8.
106. Hartlib, *Cornucopia*, pp. 15–16.
107. Hartlib, *Essay for the Advancement of Husbandrie-Learning*, p. 9.
108. Ibid., p. 9.
109. Ibid., p. 10.
110. Ibid., p. 10.
111. Ibid., p. 10.
112. Ibid., p. 11.
113. S. Hartlib, *Some Proposals Towards the Advancement of Learning*, reprinted in Webster, *Samuel Hartlib and the Advancement of Learning*, p. 179.

114. Shapin and Schaffer, *Leviathan and the Air Pump*, esp. pp. 3–22.
115. G. Boate, *Ireland's Natural History* (London, 1657), sig. A4.
116. Ibid., sig. A4 v–sig. A5 r.
117. Ibid., sig. A3.
118. P. Coughlan, 'Natural History and Historical Nature: The Project for a Natural History of Ireland', in Greengrass et al., *Samuel Hartlib*, pp. 298–317, on p. 302. For a general history of the English approach to colonizing Ireland, see J. H. Ohlmeyer, '"Civilizinge of Those Rude Partes": Colonization within Britain and Ireland, 1580s–1640s', in Canny (ed.), *OHBE, 1: Origins of Empire*, pp. 124–47; Barnard, *Cromwellian Ireland*.
119. Coughlan, 'Natural History', p. 302 (my emphasis).
120. Arnold Boate also contributed a twenty-five-page section to the second edition of Hartlib's *Legacy* in 1652, which was titled 'An Interrogatory Relating more particularly to the Husbandry and Natural History of Ireland'. This consisted of information about aspects of Irish natural environment such as fish, types of fruit trees and vegetables and how they were useful.
121. J. Dury, 'Epistle Dedicatory', in Boate, *Ireland's Natural History*, sig. A3 v–r.
122. Coughlan, 'Natural History', p. 303.
123. Poovey, *History of the Modern Fact*, p. 120.
124. W. Petty, *Political Arithmetick* (London, 1690), p. 81. On the development of political arithmetic in the eighteenth century, see P. Buck, 'People Who Counted: Political Arithmetic in the Eighteenth Century', *Isis*, 73 (1982), pp. 28–45; A. Rusnock, 'Biopolitics: Political Arithmetic in the Enlightenment', in W. Clark, J. Golinski and S. Schaffer (eds), *The Sciences in Enlightened Europe* (Chicago, IL: University of Chicago Press, 1999), pp. 49–68. In the American context, see P. C. Cohen, *A Calculating People: The Spread of Numeracy in Early America* (Chicago, IL: University of Chicago Press, 1983).
125. William Petty, *Political Arithmetick*, p. 83.
126. Ibid., p. 11.
127. Ibid., p. 79.
128. Ibid., p. 79.
129. Ibid., p. 83.
130. Ibid., p. 83.
131. Ibid., p. 77.
132. Ibid., p. 94.
133. Petty, *Political Arithmetick*, sig. A4 v.
134. Ibid., sig. A4 v.

3 Robert Boyle's Protestant Colonial Project

1. R. Boyle, *Experimenta et Observationes Physicae* (1691) in M. Hunter and E. B. Davis (eds), *The Works of Robert Boyle*, 14 vols (London: Pickering & Chatto, 2000), vol. 11, p. 377. All quotations from Boyle's work are from this edition, henceforth *Works*. All quotations from Boyle's correspondence are from M. Hunter, A. Clericuzio and L. M. Principe (eds), *The Correspondence of Robert Boyle*, 6 vols (London: Pickering & Chatto, 2001), henceforth *Corr.*

2. R. Boyle, *Some Considerations Touching the Usefulness of Experimental Natural Philosophy*, Part II (1671), *Works*, vol. 6, p. 406. Parts I and II of this work were published under slight variations on the same title. To avoid confusion I will henceforth use the abbreviation *Usefulness of Experimental Natural Philosophy* for both parts.

3. J. R. Jacob, *Robert Boyle and the English Revolution: A Study in Social and Intellectual Change* (New York: Burt Franklin & Co., 1977), p. 154. See also Jacob's more recent analysis of the relationship between trade, political economy and natural philosophy in 'The Political Economy of Science in Seventeenth-Century England', *Social Research*, 59 (1992), pp. 505–32.

4. Shapin and Schaffer, *Leviathan and the Air Pump*; S. Shapin, 'The House of Experiment in Seventeenth-Century England', *Isis*, 79 (1988), pp. 373–404; S. Shapin, *A Social History of Truth: Civility and Science in Seventeenth-Century England* (Chicago, IL: University of Chicago Press, 1994).

5. M. Hunter, *Robert Boyle (1627–91): Scrupulosity and Science* (Woodbridge: Boydell & Brewer, 2000).

6. M. Hunter (ed.), *Robert Boyle Reconsidered* (Cambridge: Cambridge University Press: 1994), p. 10.

7. Ibid., p. 9.

8. For an account of Boyle's life by himself and his friends, see M. Hunter, *Robert Boyle by Himself and his Friends, with a Fragment of William Wotton's Lost 'Life of Boyle'* (London: Pickering & Chatto, 1994).

9. See the discussion in Webster, *The Great Instauration*, pp. 61–8.

10. While living in London, Lady Ranelagh's home became the meeting point for the Irish Protestants in exile. Ibid., p. 62.

11. Ibid., p. 431.

12. In 1680, for example, he turned down the offer of the presidency of the Royal Society.

13. Webster, *The Great Instauration*, pp. 62–3.

14. J. J. O'Brien, 'Samuel Hartlib's Influence on Robert Boyle's Scientific Development Part II: Boyle in Oxford', *Annals of Science*, 21 (1965), p. 260.

15. W. Kellaway, *The New England Company 1649–1776: Missionary Society to the American Indians* (London: Longmans, 1961), p. 20n.

16. Ibid., p. 46.

17. Jacob, *Robert Boyle and the English Revolution*, p. 144.

18. J. E. McGuire, 'Boyle's Conception of Nature', *Journal of the History of Ideas*, 33 (1972), pp. 523–42, on p. 525. On Boyle and his debate with the Cambridge Platonists, see also R. Crocker, *Henry More 1614–1687: A Biography of the Cambridge Platonist* (Dordrecht: Kluwer, 2003), esp. pp. 157–60; R. A. Greene, 'Henry More and Robert Boyle on the Spirit of Nature', *Journal of the History of Ideas*, 23 (1962), pp. 451–74.

19. T. Shanahan, 'God and Nature in the Thought of Robert Boyle', *Journal of the History of Philosophy*, 26 (1988), pp. 547–69.

20. R. Boyle, *A Disquisition about the Final Causes of Natural Things: Wherein it is Inquir'd, whether, And (if at all) with what Cautions, a Naturalist should admit Them?* (1688), *Works*, vol. 11, p. 108. On Boyle and final causes, see M. J. Osler, 'From Immanent Natures to Nature as Artifice: The Reinterpretation of Final Causes in Seventeenth-Century Natural Philosophy', *The Monist*, 79 (1996), pp. 388–408.

21. M. Hunter, 'Science and Heterodoxy: An Early Modern Problem Reconsidered', in D. C. Lindberg and R. S. Westman (eds), *Reappraisals of the Scientific Revolution* (Cambridge: Cambridge University Press, 1990).

22. Boyle, *Final Causes, Works*, vol. 11, p. 108.

23. Ibid., p. 108.

24. For more of Boyle's discussion of Adam and associated ideas about paradise, knowledge and the Fall, see *Of the High Veneration Man's Intellect owes to God; Peculiarly for his*

Wisdom and Power (1685), *Works*, vol. 10, pp. 180, 182; *A Free Enquiry into the Vulgarly Receiv'd Notion of Nature; Made in an Essay Address'd to a Friend* (1686), *Works*, vol. 10, p. 494; *The Martyrdom of Theodora, and of Didymus* (1687), *Works*, vol. 11, pp. 10, 55; *Final Causes*, *Works*, vol. 11, p. 122; *The Christian Virtuoso*, Part II (1690–1), *Works*, vol. 12, pp. 499, 522, 528, 441–2; *The Christian Virtuoso*, Part I (1690–1), Appendix, *Works*, vol. 12, p. 415; *Usefulness of Experimental Natural Philosophy*, Part II, *Works*, vol. 6, p. 522; *The Excellency of Theology, Compar'd with Natural Philosophy* (1674), *Works*, vol. 8, pp. 38–9.

25. Boyle, *Usefulness of Experimental Natural Philosophy*, Part II, in *Works*, vol. 6, p. 405.
26. This was also a pseudonym for Boyle's nephew.
27. Boyle, *Usefulness of Experimental Natural Philosophy*, Part I, in *Works*, vol. 3, p. 211.
28. Ibid., vol. 3, p. 212.
29. Ibid., vol. 3, p. 206.
30. Ibid., vol. 3, p. 219.
31. Ibid., vol. 3, p. 212.
32. Ibid., vol. 3, p. 206.
33. Ibid., vol. 3, p. 206.
34. Ibid., vol. 3, p. 220.
35. Ibid., vol. 3, p. 220.
36. Ibid., vol. 3, p. 296.
37. Ibid., vol. 3, p. 296.
38. Ibid., vol. 3, p. 295.
39. Ibid., vol. 3, p. 296.
40. Ibid., vol. 3, p. 298.
41. Ibid., vol. 3, p. 229.
42. Ibid., vol. 3, p. 229.
43. Boyle, *Usefulness of Experimental Natural Philosophy*, Part II, in ibid., vol. 6, p. 425.
44. Boyle, *Usefulness of Experimental Natural Philosophy*, Part I, in ibid., vol. 3, p. 218.
45. Ibid., vol. 3, p. 212.
46. Boyle, *Usefulness of Experimental Natural Philosophy*, Part II, in ibid., vol. 6, p. 407.
47. Ibid., vol. 6, p. 407.
48. Boyle, *The Christian Virtuoso*, Part II, in ibid., vol. 12, p. 489.
49. Ibid., vol. 12, p. 444.
50. Hunter, *Robert Boyle Reconsidered*, p. 11.
51. Boyle, *Usefulness of Experimental Natural Philosophy*, Part II, *Works*, vol. 3, p. 207.
52. Ibid., vol. 3, pp. 206–7.
53. R. Iliffe, 'Foreign Bodies: Travel, Empire and the Early Royal Society of London, Part II: The Land of Experimental Knowledge', *Canadian Journal of History*, 34 (1999), p. 25.
54. Ibid., p. 25.
55. Boyle, *Usefulness of Experimental Natural Philosophy*, Part II, *Works*, vol. 6, p. 424.
56. Ibid., vol. 6, p. 425.
57. Ibid., vol. 6, p. 422.
58. Ibid., vol. 6, p. 417.
59. Boyle, *The Christian Virtuoso*, Part II, in ibid., vol. 12, pp. 444–5.
60. E. E Rich (ed.), *Minutes of the Hudson's Bay Company 1679–1684, First Part 1679–1682* (Toronto: Champlain Society, 1945), pp. 307–8.

61. See for example Robert Boyle, *Work-Diary XXXVI*, *BP*21, 285. For Boyle's *Work-Diaries*, I have cited the number of the *Work-Diary* as well as the Corresponding *Boyle Papers* (*BP*) reference. See www.bbk.ac.uk/Boyle/workdiaries/ (accessed 26 March 2008).

62. Rich (ed.), *Minutes of the Hudson's Bay Company*, pp. 307–8.

63. *BP*40, 16–46.

64. Jacob, *Robert Boyle and the English Revolution*, p. 149.

65. Robert Boyle to the Commissioners of the United Colonies in New England, 15 May 1662, *Corr.*, vol. 2, p. 20.

66. Robert Boyle to Michael Boyle, 27 May 1662, in ibid., vol. 2, p. 24.

67. John Winthrop to Robert Boyle, [1662?], in ibid., vol. 2, p. 57.

68. Ibid., vol. 2, p. 57.

69. Ibid., vol. 2, p. 57.

70. Jacob, *Robert Boyle and the English Revolution*, p. 143.

71. This is an anachronistic term, the use of which is argued against below.

72. Jacob, *Robert Boyle and the English Revolution*, p. 159.

73. Robert Boyle to Robert Thompson, 5 March 1677, *Corr.*, vol. 4, p. 436.

74. Jacob, *Robert Boyle and the English Revolution*, p. 159.

75. Quoted in R. E. W Maddison, 'Studies in the Life of Robert Boyle, F.R.S. Part I. Robert Boyle and Some of His Foreign Visitors', *Notes and Records of the Royal Society of London*, 9 (1951), p. 2.

76. Boyle quoted in M. Deacon, *Scientists and the Sea 1650–1900: A Study of Marine Science*, 2nd edn (1971; London: Ashgate, 1997), p. 119.

77. Boyle quoted in ibid., p. 119.

78. R. Boyle, *The Christian Virtuoso*, Part I, *Works*, vol. 11, p. 314.

79. Ibid., vol. 11, p. 314.

80. Ibid., vol. 11, p. 314.

81. Boyle, *Usefulness of Experimental Natural Philosophy*, Part II, *Works*, vol. 6, p. 451.

82. Ibid., vol. 6, p. 476.

83. Ibid., vol. 6, p. 435.

84. Ibid., vol. 6, p. 468.

85. Ibid., vol. 6, p. 468.

86. Ibid., vol. 6, p. 469.

87. Ibid., vol. 6, p. 469.

88. Ibid., vol. 6, p. 469.

89. Ibid., vol. 6, p. 469.

90. Ibid., vol. 6, pp. 469–70.

91. Ibid., vol. 6, p. 470.

92. A. Perez-Ramos, *Francis Bacon's Idea of Science and the Maker's Knowledge Tradition* (Oxford: Clarendon, 1988), p. 170.

93. M. Poovey, *A History of the Modern Fact: Problems of Knowledge in the Sciences of Wealth and Society* (Chicago, IL: University of Chicago Press, 1998), pp. 115–16.

94. S. Shapin, 'Pump and Circumstance: Robert Boyle's Literary Technology', *Social Studies of Science*, 14 (1984), p. 481.

95. Ibid., p. 495. On issues of reliability in the gold trade, see S. Schaffer, 'Golden Means: Assay Instruments and the Geography of Precision in the Guinea Trade', in M.-N. Bourguet, C. Licoppe and H. O. Sibum (eds), *Instruments, Travel and Science: Itineraries of Precision from the Seventeenth to the Twentieth Century* (London and New York: Routledge, 2002), pp. 20–50.

96. R. Iliffe, 'Foreign Bodies: Travel, Empire and the Early Royal Society of London, Part I: Englishmen on Tour', *Canadian Journal of History*, 33 (1998), p. 359.
97. On Stubbe, see J. R. Jacob, *Henry Stubbe, Radical Protestantism and the Early Enlightenment* (Cambridge: Cambridge University Press, 1983).
98. Boyle, *Experimenta et Observationes Physicae, Works*, vol. 11, p. 435.
99. Ibid., p. 435.
100. For an introduction to Boyle's *Work-Diaries*, see M. Hunter and C. Littleton, 'The Work-Diaries of Robert Boyle: A Newly Discovered Source and its Internet Publication', *Notes and Records of the Royal Society*, 55 (2001), pp. 373–90.
101. Boyle, *Usefulness of Experimental Natural Philosophy*, Part II, *Works*, vol. 3, p. 360.
102. Ibid., vol. 3, p. 361.
103. Robert Boyle to John Winthrop, 21 April 1664, *Corr.*, vol. 2, p. 268.
104. Robert Boyle to John Winthrop, 17 March 1665, in ibid., vol. 2, p. 462.
105. John Winthrop to Robert Boyle, 27 July 1662, in ibid., vol. 2, p. 33.
106. Robert Boyle, *Cosmical Suspicions, Works*, vol. 6, p. 308.
107. Ibid., vol. 6, p. 308.
108. Boyle, *General History of the Air, Works*, vol. 12, p. 102.
109. Ibid., vol. 12, p. 101.
110. Ibid., vol. 12, p. 102.
111. Ibid., vol. 12, p. 108.
112. Boyle, *Work-Diary XXXVI, BP*21, 255.
113. Ibid., 261.
114. Boyle, *Work-Diary XXI, BP*27, 45.
115. Boyle, *Work-Diary XXXVI, BP*21, 293.
116. Ibid., 293.
117. Ibid., 294.
118. Ibid., 296.
119. Boyle, *The Christian Virtuoso, Works*, vol. 12, p. 443.
120. Boyle, *General History of the Air, Works*, vol. 12, p. 112.
121. Ibid., vol. 12, p. 148.

4 The Royal Society and the Atlantic World

1. Henry Oldenburg to John Winthrop, 13 October 1667, in *The Correspondence of Henry Oldenburg*, ed. and trans. A. R. Hall and M. B. Hall, 13 vols (Madison, WI: University of Wisconsin Press, 1965), vol. 3, p. 525.
2. *Philosophical Transactions of the Royal Society*, 16:180 (1686), p. 37. References are given as volume: number (year), page. Note that some volumes do not have issue numbers.
3. Robert Hooke quoted in M. Hunter, *Establishing the New Science: The Experience of the Early Royal Society* (Woodbridge: Boydell & Brewer, 1989), p. 138.
4. Henry Oldenburg to René François Sluse, 23 October 1667, *Oldenburg Correspondence*, vol. 3, p. 537.
5. Ibid.
6. Ibid., p. 242.
7. Hunter, *Establishing the New Science*, 1.
8. The Society included Catholic Fellows such as Kenelm Digby.
9. On Oldenburg, see M. B. Hall, *Henry Oldenburg: Shaping the Royal Society* (Oxford: Oxford University Press, 2002).

10. Hunter, *Establishing the New Science*, 67.
11. The point is made by Frances Yates in *The Rosicrucian Enlightenment* (1972; London: Routledge, 2002), pp. 243–4.
12. M. Poovey, *A History of the Modern Fact: Problems of Knowledge in the Sciences of Wealth and Society* (Chicago, IL: University of Chicago Press, 1998), p. 110.
13. Ibid., p. 111.
14. T. Sprat, *The History of the Royal Society of London, For the Improving of Natural Knowledge* (London, 1667), p. 348.
15. Ibid., p. 349.
16. Ibid., pp. 349–50.
17. Ibid., p. 351.
18. E. Ashmole, *Theatrum Chemicum Britannium* (London, 1651), p. 444.
19. Ibid., p. 445.
20. J. Glanvill, 'Modern Improvements of Knowledge', *Essays on Several Important Subjects in Philosophy and Religion* (London, 1676), pp. 46-7.
21. Ibid., p. 46.
22. Henry Oldenburg to Olaus Rudbeck, 8 January 1668/9, *Oldenburg Correspondence*, vol. 5, p. 326.
23. Henry Oldenburg to ? Mentzel, 12 February 1668/9, in ibid., vol. 5, p. 394.
24. Henry Oldenburg to Richard Norwood, 10 February 1667/8, in ibid., vol. 4, p. 168.
25. On the context of the reform in natural history in early modern Europe, see P. Findlen, 'Natural History', in K. Park and L. Daston (eds), *The Cambridge History of Science, Volume 3: Early Modern Science* (Cambridge: Cambridge University Press: 2006), pp. 437–71.
26. Sprat, *History of the Royal Society of London*, p. 155.
27. Ibid., p. 156.
28. *Philosophical Transactions*, 17:203 (1693), p. 978.
29. Ibid., 19:220 (1695–7), pp. 225–8.
30. Ibid., 5:57 (1670), p. 1151.
31. John Locke to Henry Oldenburg, 20 May 1675, *Oldenburg Correspondence*, vol. 11, p. 2667.
32. Martin Lister to Henry Oldenburg, 27 June 1673, in ibid., vol. 11, p. 373.
33. *Philosophical Transactions*, 7:83 (1672), p. 4078.
34. Hans Sloane, *Voyage to the Islands of Madera, Barbados, Nieves, S Christophers and Jamaica, with the natural history of the last of those islands; to which is prefixed an introduction, wherein is an account of the inhabitants, air, waters, diseases, trade &c.* (London, 1707–25).
35. Henry Oldenburg to John Winthrop, 11 April 1671, *Oldenburg Correspondence*, vol. 7, p. 569.
36. Henry Oldenburg to Richard Norwood, 24 October 1666, ibid., vol. 3, p. 276.
37. Richard Stafford to Henry Oldenburg, 16 July 1668, ibid., vol. 4, p. 552.
38. *Philosophical Transactions*, 1:8 (1665–6), p. 141.
39. Ibid., 1:8 (1665–6), p. 141.
40. Ibid., 1:11 (1666), p. 188.
41. Ibid., 1:11 (1666), p. 187.
42. Ibid., 11:23 (1666–7), pp. 420–1.
43. Ibid., 11:23 (1666–7), p. 422.
44. Ibid., 3:33 (1667–8), p. 634.

45. Ibid., 3:33 (1667/8), p. 634.
46. Ibid., 3:33 (1667–8), p. 634.
47. Ibid., 3:33 (1667–8), pp. 634–9.
48. H. Stubbe, *Legends, No Histories, or A Specimen of Some Animadversions Upon the History of the Royal Society* (London, 1670), sig. A1 r.
49. H. Stubbe, *The Lord Bacons Relation to the Sweating-Sickness Examined, in a Reply to George Thomson* (London, 1671), sig. LI 2.
50. Stubbe, *Legends, No Histories*, sig. A1 r.
51. John Winthrop to Henry Oldenburg, 4 October 1669, *Oldenburg Correspondence*, vol. 6, pp. 256–7.
52. Hooke and Sprat, quoted in Hunter, *Establishing the New Science*, p. 138.
53. See S. Atran, *The Cognitive Foundations of Natural History: Towards an Anthropology of Science* (Cambridge: Cambridge University Press, 1990).
54. M. Swann, *Curiosities and Texts: the Culture of Collecting in Early Modern England* (Philadelphia, PA: University of Pennsylvania Press, 2001), p. 18.
55. M. Foucault, *The Order of Things: The Archaeology of the Human Sciences* (New York: Pantheon, 1970).
56. *Calendar of State Papers Colonial Series, America and the West Indies* (London: Longman, 1860), 1, no. 11, 1637.
57. N. Grew, *Musaeum Regalis Societatis, or, A Catalogue and Description of the Natural and Artificial Rarities Belonging to the Royal Society and Preserved at Gresham Colledge* (London, 1686).
58. Ibid., sig. A4 r.
59. Ibid., Preface, sig. A.
60. C. F. Feest, 'The Collecting of American Indian Artifacts in Europe 1493–1750', in K. O. Kupperman (ed.), *America in European Consciousness 1493–1750* (Chapel Hill, NC: University of North Carolina Press, 1995), pp. 324–50, on p. 336.
61. A. MacGregor, *Tradescant's Rarities: Essays on the Foundation of the Ashmolean Museum, 1683, with a Catalogue of the Surviving Early Collections* (Oxford: Oxford University Press, 1983), pp. 108–39.
62. On Ashmole, see A. G. MacGregor and A. J Turner, 'The Ashmolean Museum', in L. S. Sutherland and L. G. Mitchell (eds), *The Eighteenth Century, History of the University of Oxford, Volume 5* (Oxford: Clarendon Press, 1986), pp. 639–58; C. H. Josten, *Elias Ashmole: His Autobiographical and Historical Notes, His Correspondence, and Other Contemporary Sources Relating to His Life and Work* (Oxford: Oxford University Press, 1966).
63. 'Museum', *Oxford English Dictionary* (Oxford: 2003), electronic edition.
64. 'Repository', ibid.
65. J. Evelyn, *Diary*, ed. E. S. de Beer, 6 vols (Oxford: Clarendon Press, 1955), 6 February 1645, vol. 2, pp. 330–1.
66. Henry Oldenburg to Robert Boyle, 27 January 1665/6, *Oldenburg Correspondence*, vol. 3, p. 32.
67. Henry Oldenburg to Robert Boyle, 24 February 1665/6, in ibid., vol. 3, p. 45.
68. Henry Oldenburg to Benjamin Lannoy, 21 November 1668, in ibid., vol. 5, p. 200.
69. Henry Oldenburg to Stephen Flower, 21 November 1668, in ibid., vol. 5, p. 201.
70. *Philosophical Transactions*, 22 (1700–1), p. 579.
71. Henry Oldenburg to Lord Williamson, ? September 1673, *Oldenburg Correspondence*, vol. 10, p. 176.

72. Henry Oldenburg to Richard Norwood, 24 October 1666, in ibid., vol. 3, p. 276.
73. Henry Oldenburg to Martin Lister, 13 December 1673, in ibid., vol. 10, p. 407.
74. Henry Oldenburg to John Winthrop, 26 March 1664, in ibid., vol. 2, p. 149.
75. Ibid., vol. 2, p. 150.
76. Henry Oldenburg to John Winthrop, 26 March 1670, in ibid., vol. 6, p. 594.
77. Henry Oldenburg to Robert Southwell, 27 May 1669, in ibid., vol. 5, p. 564.
78. William Ball to Henry Oldenburg, 14 April 1666, in ibid., vol. 3, p. 90.
79. Silas Taylor to Henry Oldenburg, 17 April 1666, in ibid., vol. 3, p. 94.
80. Richard Stafford to Henry Oldenburg, 16 July 1668, in ibid., vol. 4, p. 552.
81. John Winthrop to Henry Oldenburg, 4 October 1669, in ibid., vol. 6, p. 253.
82. John Winthrop to Henry Oldenburg, 26 August 1670, in ibid., vol. 7, p. 143.
83. John Winthrop to Henry Oldenburg, 25 September 1672, in ibid., vol. 9, p. 257.
84. Evelyn, *Diary*, 27 August–14 September 1662, vol. 3, p. 334.
85. Ibid., 17 May–26 July 1668, vol. 3, p. 511.
86. Ibid., 23 September–4 November 1670, vol. 3, p. 563.
87. *Philosophical Transactions*, 10 (1675), p. 255.
88. Ibid., 11:127 (1676), p. 647.
89. Ibid., 20 (1698), sig. A1 v.
90. *Philosophical Transactions*, 5:57 (1670), p. 1151.

5 John Locke's Language of Empire

1. B. Arneil, 'Trade, Plantations and Property: John Locke and the Economic Defence of Colonialism', *Journal of the History of Ideas*, 55 (1994), pp. 591–609; B. Arneil *John Locke and America: The Defence of English Colonialism* (Oxford: Oxford University Press, 1996); H. Lebovic, 'The Uses of America in Locke's Second Treatise of Government', *Journal of the History of Ideas*, 47 (1986), pp. 567–81. J. Tully, 'Rediscovering America: The *Two Treatises* and Aboriginal Rights', Chapter 5 in J. Tully, *An Approach to Political Philosophy: Locke in Contexts* (Cambridge: Cambridge University Press, 1993), pp. 137–76.

2. A. Pagden, 'The Struggle for Legitimacy and the Image of the Empire in the Atlantic to *c.* 1700', in N. Canny (ed.), *The Oxford History of the British Empire, Volume 1: The Origins of Empire* (Oxford: Oxford University Press, 1999), pp. 42–7.

3. D. Ivison, 'Locke, Liberalism and Empire', in P. Anstey (ed.), *The Philosophy of John Locke: New Perspectives* (London: Routledge, 2003), pp. 86–105.

4. D. Armitage, 'John Locke, Carolina and the *Two Treatises of Government*', *Political Theory*, 32 (2004), pp. 602–27; D. Armitage, *The Ideological Origins of the British Empire* (Cambridge: Cambridge University Press, 2000), pp. 96–9, 165–6.

5. J. Locke, *Essay Concerning Human Understanding*, ed. P. H. Nidditch (Oxford: Clarendon Press, 1975), Book 4, XII, §11. Henceforth *ECHU*. References are given as book, chapter, section.

6. Ibid., § 10.

7. Ibid., § 11.

8. Colonialism is an anachronistic term, but I use it because the historians to whom I refer use it to describe their work. See for example D. Armitage, 'John Locke, Carolina and the Two Treatises of Government', p. 602. Armitage lists 'colonialism' as one of the keywords for his article.

9.   J. Locke, *Two Treatises of Government*, ed. P. Laslett (1960; Cambridge: Cambridge University Press, 1963), 1, III, §27. References to the *Two Treatises* are given Treatise number, chapter, section number.

10. Ibid., 1, IV, §40.

11. Ibid., 1, IV, §41.

12. Coke was the author of *A Discourse of Trade in Two Parts: The Reason of the Decay of the Strength, Wealth and Trade of England* (London, 1670), cited in Arneil, 'Trade, Plantations and Property', p. 595.

13. Ibid., p. 597. This key argument is also explained in Arneil, *John Locke and America*, pp. 90–117.

14. Arneil, 'Trade, Plantations and Property', p. 592.

15. Ibid., p. 602.

16. Ibid., p. 603.

17. Armitage, 'John Locke, Carolina and the *Two Treatises of Government*', p. 617.

18. Ibid., p. 618.

19. Ibid., p. 618.

20. Tully, *Locke in Contexts*, p. 151.

21. J. Tully, *A Discourse on Property: John Locke and His Adversaries* (Cambridge: Cambridge University Press, 1980), p. 174.

22. Locke quoted in ibid., p. 60.

23. Locke quoted in ibid., 60.

24. Ibid., 173.

25. J. Dunn, *The Political Thought of John Locke* (Cambridge: Cambridge University Press, 1968); J. Dunn, 'From Applied Theology to Social Analysis' in *Rethinking Modern Political Theory Essays 1979–83* (Cambridge: Cambridge University Press, 1985). More recently there has been a debate ignited by Jeremy Waldron's 1999 Carlyle Lectures, though this debate occurs primarily in the context of political theory. See the contributions to the symposium 'God, Locke and Equality' in *Review of Politics*, 67 (2005), *passim*; J. Waldron, *God, Locke and Equality: Christian Foundations of Locke's Political Thought* (Cambridge: Cambridge University Press, 2002).

26. P. Harrison, '"Fill the Earth and Subdue It"', p. 18. See also P. Harrison, 'Subduing the Earth: Genesis 1, Early Modern Science, and the Exploration of Nature', *Journal of Religion*, 79 (1999), pp. 86–109.

27. W. Glausser, 'Three Approaches to Locke and the Slave Trade', *Journal of the History of Ideas*, p. 51 (1990), p. 200. Locke sold his shares in 1675. See M. Cranston, *John Locke: A Biography* (New York: Macmillan, 1957), p. 115.

28. Glausser, 'Three Approaches', p. 201.

29. Ibid., p. 201.

30. Quoted in R. Ashcraft, 'Political Theory and Political Reform: John Locke's Essay on Virginia', *Western Political Quarterly*, 22 (1969), p. 744.

31. J. Locke 'Understanding', in *Political Essays*, ed. Goldie, p. 261.

32. Ibid., p. 260.

33. Ibid., p. 260.

34. K. Park and L. Daston, 'Introduction', in K. Park and L. Daston (eds), *Cambridge History of Science, Volume 3: Early Modern Science* (Cambridge: Cambridge University Press, 2006), p. 14.

35. M. Goldie, introduction to Locke's essay on 'Understanding', in *Political Essays*, ed. Goldie, p. 260.

36. Locke, 'Understanding', in ibid., p. 262.
37. Ibid., p. 261.
38. Ibid., p. 260.
39. M. Ayers, *Locke: Epistemology and Ontology* (1993; London: Routledge, 1999), p. 13.
40. Locke, *ECHU*, Book 4, XII, §10.
41. J. Locke, *Mr. Locke's Reply to the Right Reverend the Lord Bishop of Worcester* (London: 1697), p. 75.
42. See especially Locke's corrections in Locke MS c37 (Bodleian Library).
43. Locke, 'Understanding', pp. 264–5.
44. J. Locke, 'Homo ante et post Lapsum', in *Political Essays*, ed. Goldie, p. 320. The essay was written in 1693.
45. Ibid., p. 320.
46. Ibid., p. 321.
47. Ibid., p. 321.
48. Locke, *Two Treatises of Government*, 2, VI, §56.
49. *ECHU*, Book 3, VI, §44.
50. Ibid., §44.
51. Ibid., §46.
52. Ibid., §47.
53. H. Aarslef, *From Locke to Saussure: Essays on the Study of Language and Intellectual History* (Minneapolis, MN: University of Minnesota Press, 1982), p. 26.
54. Ibid., pp. 25–7.
55. Locke, *ECHU*, Book 4, XII, §10.
56. Locke, *Reply to the Bishop of Worcester*, p. 266.
57. Ibid., p. 51.
58. Locke, *ECHU*, Book 4, XII, §11.
59. J. Locke, 'Some Thoughts Concerning Reading and Study for a Gentleman', in *Political Essays*, ed. Goldie, p. 353.
60. Ibid., p. 353. All these authors with the exception of Roe appear in Locke's Library Catalogue, J. Harrison and P. Laslett (eds), *The Library of John Locke* (1965; Oxford: Oxford University Press, 1971).
61. Locke, 'Some Thoughts Concerning Reading', in *Political Essays*, ed. Goldie, p. 353.
62. J. Locke, 'Study', in *Political Essays*, ed. Goldie, p. 367.
63. Locke, *Two Treatises of Government*, 1, III, §27.
64. Ibid., 1, IX, §86.
65. Ibid., 1, IV, §40.
66. Ibid., 1, XI, §112.
67. Locke quoted in Tully, *Discourse on Property*, p. 36.
68. Locke, *Two Treatises of Government*, 1, XI, §113.
69. Ibid., 1, XI, §115.
70. Ibid., 1, XI, §116.
71. Tully, *Discourse on Property*, p. 61.
72. Locke, 'Understanding', in *Political Essays*, ed. Goldie, p. 264.
73. Ibid., p. 265.
74. J. Locke, *Some Thoughts Concerning Education*, in *The Educational Writings of John Locke*, ed. J. L. Axtell (Cambridge: Cambridge University Press, 1968), p. 302.
75. Ibid., p. 301.
76. Ibid., p. 305.

77. Ibid., p. 306.
78. Ibid., p. 306.
79. Locke MS d9.
80. *Philosophical Transactions of the Royal Society*, 24:298 (1705), pp. 1917–37. Reference is given volume: number (year), page number. Locke published five letters or memoranda in the *Philosophical Transactions*.
81. 'Extracts by Locke from Letters of John Stewart concerning Carolina', 1690, Locke MS c30, fol. 31.
82. Ibid.
83. Ibid.
84. Locke MS e9, fol. 39.
85. Ibid.
86. John Locke to Henry Oldenburg, 20 May 1675, in, *Oldenburg Correspondence*, vol. 11, p. 2667.
87. Ibid.
88. *Philosophical Transactions of the Royal Society*, 10:114 (1675), p. 312.
89. Dr Henry Woodward to Locke, Westo [Georgia], 12 November 1675, in *John Locke: Selected Correspondence*, ed. M. Goldie (Oxford: Oxford University Press, 2002), letter 305, p. 52.
90. Sir Peter Colleton to John Locke, 12 August 1673, in *Selected Correspondence*, ed. Goldie, letter 275, p. 45.
91. Ibid., p. 45, n. 10.
92. Major Walker was related to Colleton and was probably James Walker of Prince Rupert's Barbados Regiment of Dragoons. Ibid., p. 45, n. 11.
93. Ibid., pp. 45–6.
94. Richard Lilburne to John Locke, 12 August 1675, in *The Correspondence of John Locke*, ed. E. S de Beer, 8 vols (Oxford: Clarendon Press, 1976–89), vol. 1, letter 300, p. 424.
95. Sir Peter Colleton to John Locke, Barbados, *c.* October 1673, in *Selected Correspondence*, ed. Goldie, letter 279, p. 47.
96. Ibid., p. 48.
97. Sir Peter Colleton to John Locke, [early summer 1671?], in *Correspondence*, ed. de Beer, vol. 1, letter 254, p. 355.
98. Ibid., pp. 355–6.
99. Ibid., p. 355, n. 1.
100. Richard Lilburne to John Locke, 6 August 1674, in *Correspondence*, ed. de Beer, vol. 1, letter 290, pp. 406–7.
101. Ibid., p. 407. As de Beer points out, the oil of soldier crabs is mentioned by Lionel Wafer in *A New Voyage and Description of the Isthmus of America* (London, 1704).
102. Richard Lilburne to John Locke, 12 August 1675, in *Correspondence*, ed. de Beer, vol. 1, letter 300, p. 424.
103. Ibid., n. 4
104. Ibid., p. 424.
105. Ibid., p. 425.
106. Ibid., p. 425.
107. Ibid., p. 425.
108. Isaac Rush to John Locke, 19 August 1675, in *Correspondence*, ed. de Beer, vol. 1, letter 301, p. 427.

109. Dr Henry Woodward to Locke, 12 November 1675, in *Selected Correspondence*, ed. Goldie, letter 305, p. 51.
110. Ibid., p. 51.
111. See Harrison and Laslett (eds), *Library of John Locke*, p. 97.
112. Dr Henrry Woodward to Locke, 12 November 1675, in *Selected Correspondence*, ed. Goldie, letter 305, p. 51.
113. Joseph West to John Locke, 4 September 1676, in *Correspondence*, ed. de Beer, vol. 1, letter 318, p. 457.
114. Ibid.
115. Locke MS c.33, fol. 11.
116. Locke MS c37, fol. 57.
117. Ibid., fol. 69.
118. J. Locke, 'Advertisement of the Publisher to the Reader', in R. Boyle, *The General History of the Air, Design'd and Begun by the Honble Robert Boyle Esq.* (London, 1692), p. iii.
119. Ibid, p. iii.
120. Locke, *Two Treatises of Government*, 1, IV, §33.
121. Ibid., 1, IV, §33.
122. Ibid., 1, IV, §39.
123. Ibid., 1, IV, §40.
124. Ibid., 1, IV, §41.
125. Ibid., 2, V, §26.
126. Ibid., 2, V, §27.
127. D. Armitage, 'John Locke, Carolina and the *Two Treatises of Government*', p. 617.
128. Locke, *Two Treatises of Government*, 2, V, §31; see also 1, §40, 22.
129. Ibid., 2, V, §32.
130. Ibid., 2, V, §35.
131. Ibid.
132. J. Locke, *Some Considerations of the Consequences of the Lowering of Interest and Raising the Value of Money* (London, 1691), p. 15.
133. Ibid., p. 99.
134. J. Locke to John Cary, Oates, 2 May 1696, in *Selected Correspondence*, ed. Goldie, letter 2064, p. 227.
135. J. Locke, 'Trade' in *Political Essays*, ed. Goldie, p. 221.
136. Ibid., p. 222.
137. Locke MS c30, fol. 67.
138. Ibid.
139. Locke MS c30, fol. 84.
140. Ibid.
141. Ibid.
142. Locke MS c30, fol. 59.
143. Locke MS e9, fol. 17.
144. J. Locke, *Some Thoughts Concerning Education*, in *Educational Writings*, ed. Axtell, p. 314.
145. Ibid., pp. 315–16.
146. Ibid., p. 316.
147. Ibid., p. 317.
148. J. Locke to Edward Clarke, 6 February 1688, in *Educational Writings*, ed. Axtell, p. 381.
149. Tully, *Locke in Contexts*, pp. 179–42.

150. J. Locke, 'An Essay on the Poor Law', in *Political Essays*, ed. Goldie, p. 183.
151. Ibid., p. 184.
152. J. Locke, 'For a General Naturalisation', in *Political Essays*, ed. Goldie, p. 322.
153. Ibid., p. 323.
154. Ibid., p. 325.
155. J. Locke, 'Labour', in *Political Essays*, ed. Goldie, p. 326.
156. Ibid.
157. Ibid.
158. Ibid., p. 328.
159. Ibid.
160. Ibid.

Conclusion

1.   W. Jarvis, quoted in D. G. Bell, 'Was Amerindian Dispossession Lawful? The Response of 19th-Century Maritime Intellectuals', *Dalhousie Law Journal*, 23 (2000), p. 176. I would like to thank Jon Penney for drawing my attention to this discussion.
2.   Ibid., p. 177.
3.   B. Buchan and M. Heath, 'Savagery and Civilization: From *Terra Nullius* to the "Tide of History"', *Ethnicities*, 6 (2006), p. 10.
4.   Justice Deane and Justice Gaudron in *Mabo and Others* v. *Queensland* (no. 2) (1992), p. 6.
5.   Peter Harrison argues that the second half of the seventeenth century 'witnessed the beginnings of a blurring of boundaries between these two approaches': P. Harrison, '"Fill the Earth and Subdue It"', p. 23.

WORKS CITED

Primary Sources

Ashmole, E., *Theatrum Chemicum Britannium* (London, 1651).

Bacon, F., *The Works of Francis Bacon*, ed. J. Spedding, R. L. Ellis and D. D. Heath, 14 vols (London: Longman, 1857–74).

—, 'The Masculine Birth of Time', reprinted in B. Farrington, *The Philosophy of Francis Bacon: An Essay on its Development from 1603–1609* (Liverpool: Liverpool University Press, 1964).

—, *Collected Works of Francis Bacon*, ed. J. Spedding, R. L. Ellis and D. D. Heath, with a new introduction by G. Rees, 7 vols (London: Routledge/Thoemmes, 1996).

—, 'De Fluxu et Refluxu Maris', in *Philosophical Studies c. 1611–1619*, in *Oxford Francis Bacon, Volume 6*, ed. G. Rees (Oxford: Clarendon Press, 1996).

—, *The Advancement of Learning*, in *The Oxford Francis Bacon, Volume 4*, ed. M. Kiernan (Oxford: Clarendon Press, 2000).

—, *The New Organon*, ed. L. Jardine and M. Silverthorne (Cambridge: Cambridge University Press, 2000).

—, 'Of Empire', in *The Essayes or Counsels, Civill and Moral*, ed. M. Kiernan, in *Oxford Francis Bacon, Volume 15* (Oxford: Clarendon Press, 2000).

—, *Francis Bacon: The Major Works*, ed. B. Vickers (1996; Oxford: Oxford University Press, 2002).

—, *The Instauratio Magna, Part II: Novum Organon and Associated Texts, The Oxford Francis Bacon Volume 11*, ed. G. Rees, and M. Wakely (Oxford: Clarendon Press, 2004).

Boate, G., *Ireland's Natural History* (London, 1657).

Boyle, R., *Work-Diaries*. Digitized online by the Robert Boyle Project, www.bbk.ac.uk/Boyle/workdiaries/.

—, *The Correspondence of Robert Boyle*, ed. M. Hunter, A. Clericuzio and L. M. Principe, 6 vols (London: Pickering & Chatto, 2001).

—, *The Works of Robert Boyle*, ed. M. Hunter, A. Clericuzio, L. M. Principe and E. B. Davis, 14 vols (London: Pickering & Chatto, 2000).

Bullock, W., *Virginia Partially Examined, and Left to Public View, to be Considered by all Judicious and Honest Men* (London, 1649).

Calendar of State Papers Colonial Series, America and the West Indies (London: Longman, 1860).

Cicero, *De Officiis*, ed. M. T. Griffin and E. M Atkins (Cambridge: Cambridge University Press, 1996).

Evelyn, J., *Elysium Britannicum*, British Library: BL, Add. MSS 78342–78344.

—, *Diary*, ed. E. S. de Beer, 6 vols (Oxford: Clarendon Press, 1955).

Glanvill, J., 'Modern Improvements of Knowledge', *Essays on Several Important Subjects in Philosophy and Religion* (London, 1676).

Grew, N., *Musaeum Regalis Societatis, or, A Catalogue and Description of the Natural and Artificial Rarities Belonging to the Royal Society and Preserved at Gresham Colledge* (London, 1686).

The Hartlib Papers, 2nd edn (Sheffield: HROnline, 2002).

Hartlib, S., *An Essay for Advancement of Husbandry-Learning: or Propositions for the Erecting Colledge of Husbandry* (London, 1651).

—, *Samuel Hartlib his Legacie or An Enlargement of the Discourse of Husbandry Used in Brabant and Flaunders; Wherein are Bequeathed to the Common-Wealth of England More Outlandish and Domestick Experiments and Secrets in Reference to Universall Husbandry* (London, 1651).

—, *Cornucopia: A Miscellanium of Lucriferous and most Fructiferous Experiments, Observations, and Discoveries, Immethodically Distributed; to be Really Demonstrated and Communicated in All Sincerity* (London, 1652).

—, *The Commonwealth of Bees* (London, 1655).

—, *The Compleat Husband-Man: or, A Discourse of the Whole Art of Husbandry both Forraign and Domestick* (London, 1659).

King James Bible (Oxford: Oxford University Press, 1998).

Locke, J., Bodleian Library: MSS Locke.

—, *Some Considerations of the Consequences of the Lowering of Interest, and Raising the Value of Money* (London, 1691).

—, 'Advertisement of the Publisher to the Reader', in R. Boyle, *The General History of the Air, Design'd and Begun by the Honble Robert Boyle Esq* (London, 1692).

—, *Mr. Locke's Reply to the Right Reverend the Lord Bishop of Worcester* (London, 1697).

—, *Two Treatises of* Government, ed. P. Laslett (1960; Cambridge: Cambridge University Press, 1963).

—, *The Educational Writings of John Locke*, ed. J. L. Axtell (Cambridge: Cambridge University Press, 1968).

—, *Essay Concerning Human Understanding*, ed. P. H. Nidditch (Oxford: Clarendon Press, 1975).

—, *The Correspondence of John Locke*, ed. E. S de Beer, 8 vols (Oxford: Clarendon Press, 1976–89)

—, *Locke: Political Essays*, ed. M. Goldie (Cambridge: Cambridge University Press, 1997).

—, *John Locke: Selected Correspondence*, ed. M. Goldie (Oxford: Oxford University Press, 2002).

Oldenburg, H., *The Correspondence of Henry Oldenburg*, ed. and trans. Hall, A. R. and Hall, M. B., 13 vols (Madison, WI: University of Wisconsin Press, 1965).

Petty, W., *Political Arithmetick* (London, 1690).

—, *The Political Anatomy of Ireland* (London, 1691).

Philosophical Transactions of the Royal Society of London.

Rawley, W., *Resuscitatio, or Bringing into Publick Light Severall Pieces of the Works, Civil, Historical, Philosophical, and Theological, Hitherto Sleeping, of the Right Honourable Francis Bacon* (London, 1657).

Sallust, *War with Catiline*, trans. P. McGushin (Bristol: Bristol Classical Press, 1980).

Sloane, H., *Voyage to the Islands of Madera, Barbados, Nieves, S Christophers and Jamaica, with the natural history of the last of those islands; to which is prefixed an introduction, wherein is an account of the inhabitants, air, waters, diseases, trade &c* (London, 1707–25).

Speed, A., *Adam out of Eden, or an Abstract of Divers excellent Experiments Touching the Advancement of Husbandry* (London, 1659).

Sprat, T., *The History of the Royal Society of London, For the Improving of Natural Knowledge* (London, 1667).

Stubbe, H., *Legends, No Histories, or A Specimen of Some Animadversions Upon The History of the Royal Society* (London, 1670).

—, *The Lord Bacons Relation to the Sweating-Sickness Examined, in a reply to George Thomson* (London, 1671).

Taylor, E. G. R., *The Original Writings and Correspondence of the two Richard Hakluyts*, 2 vols (London: Hakluyt Society, 1935).

Secondary Sources

Aarslef, H., *From Locke to Saussure: Essays on the Study of Language and Intellectual History* (Minneapolis, MN: University of Minnesota Press, 1982).

Abraham, G. A., 'Misunderstanding the Merton Thesis: A Boundary Dispute between History and Sociology', *Isis*, 74 (1983), pp. 368–87.

Alam, M. A., 'Science and Imperialism: What is Science?', *Race & Class*, 19 (1978), pp. 241–51.

Albanese, D., 'New Atlantis and the Uses of Utopia', *English Literary History*, 57 (1990), pp. 503–28.

Allan, M., *The Tradescants: Their Plants, Gardens and Museum* (London: Michael Joseph, 1964).

Anderson, V. D., *New England's Generation: The Great Migration and the Formation of Society and Culture in the Seventeenth Century* (Cambridge: Cambridge University Press, 1991).

—, 'New England in the Seventeenth Century', in Canny (ed.) *OHBE 1: Origins of Empire*, pp. 193–217.

Andrews, C. M., *The Colonial Period of American History*, 4 vols (New Haven, CT: Yale University Press, 1934–8).

Anstey, P., *The Philosophy of Robert Boyle* (London: Routledge, 2000).

— (ed.), *The Philosophy of John Locke: New Perspectives* (London: Routledge, 2003).

—, 'Hartlib and Starkey Rekindled', *Metascience*, 13 (2004), pp. 112–15.

Armitage, D., *The Ideological Origins of the British Empire* (Cambridge: Cambridge University Press, 2000).

—, 'John Locke, Carolina and the Two Treatises of Government', *Political Theory*, 32 (2004), pp. 602–27.

Armytage, W. H. G., 'The Early Utopists and Science in England', *Annals of Science*, 12 (1956), pp. 247–54.

Arneil, B., 'Trade, Plantations and Property: John Locke and the Economic Defence of Colonialism', *Journal of the History of Ideas*, 55 (1994), pp. 591–609.

—, *John Locke and America: The Defence of English Colonialism* (Oxford: Oxford University Press, 1996).

Ashcraft, R., 'Political Theory and Political Reform: John Locke's Essay on Virginia', *Western Political Quarterly*, 22 (1969), pp. 742–58.

Atran, S., *The Cognitive Foundations of Natural History: Towards an Anthropology of Science* (Cambridge: Cambridge University Press 1990).

Ayers, M., *Locke: Epistemology and Ontology* (1993; London: Routledge, 1999).

Aylmer, G. E., 'Navy, State, Trade, and Empire', in Canny (ed.) *OHBE 1: Origins of Empire*, pp. 467–80.

Barnard, T. C., *Cromwellian Ireland: English Government and Reform in Ireland, 1649–1660* (Oxford: Oxford University Press, 1975).

Barrow, T. C., *Trade and Empire: The British Customs Service in Colonial America, 1660–1775* (Cambridge, MA: Harvard University Press, 1967).

Beckles, H. McD., '"The Hub of Empire": The Caribbean and Britain in the Seventeenth Century', in Canny (ed.) *OHBE 1: Origins of Empire*, pp. 218–40.

Bell, D. G., 'Was Amerindian Dispossession Lawful? The Response of 19th-Century Maritime Intellectuals', *Dalhousie Law Journal*, 23 (2000), pp. 168–82.

Bennett, J. and Mandlebrote, S., *The Garden, the Ark, the Tower, the Temple: Biblical Metaphors of Knowledge in Early Modern Europe* (Oxford: Bodleian Library, 1998).

Boesky, A., *Founding Fictions: Utopias in Early Modern England* (Athens: University of Georgia Press, 1996).

—, 'Bacon's *New Atlantis* and the Laboratory of Prose', in E. Fowler and R. Greene (eds) *The Project of Prose in Early Modern Europe and the New World* (Cambridge: Cambridge University Press, 1997), pp. 138–53.

Braddick, M., 'The English Government, War, Trade and Settlement 1625–1688', in Canny (ed.) *OHBE 1: Origins of Empire*, pp. 286–308.

Breen, T. H., *Puritans and Adventurers: Change and Persistence in Early America* (New York: Oxford University Press, 1980).

Brenner, R., *Merchants and Revolution: Commercial Change, Political Conflict and London's Overseas Traders, 1550–1650* (Princeton, NJ: Princeton University Press, 1993).

Brockway, L. H., *Science and Colonial Expansion: The Role of the British Royal Botanical Garden* (New York: Academic Press, 1979).

Buchan, B. and M. Heath, 'Savagery and Civilization: From *Terra Nullius* to the "Tide of History"', *Ethnicities*, 6 (2006), pp. 5–26.

Buck, P., 'People Who Counted: Political Arithmetic in the Eighteenth Century', *Isis*, 73 (1982), pp. 28–45;

Canny, N. (ed.), *The Oxford History of the British Empire (OHBE), Volume 1: The Origins of Empire* (Oxford: Oxford University Press, 1999).

—, *Making Ireland British 1580–1650* (Oxford: Oxford University Press, 2001).

Capkova, D., 'Comenius and His Ideals: Escape from the Labyrinth', in Greengrass et al., *Samuel Hartlib*, pp. 75–91.

Carey, D., 'Compiling Nature's History: Travellers and Travel Narratives in the Early Royal Society', *Annals of Science*, 54 (1997), pp. 269–92.

Chaplin, J., *Subject Matter: Technology, the Body and Science on the Anglo-American Frontier, 1500–1676* (Cambridge, MA: Harvard University Press, 2001).

Clericuzio, A., 'New Light on Benjamin Worsley's Natural Philosophy', in Greengrass et.al., *Samuel Hartlib*, pp. 236–46.

Cogley, R. W., *John Eliot's Mission to the Indians before King Philip's War* (Cambridge, MA: Harvard University Press, 1999).

Cohen, I. B. (ed.), *Puritanism and the Rise of Modern Science: The Merton Thesis* (New Brunswick and London: Rutgers University Press, 1990).

Cohen, J., *Be Fertile and Increase, Fill the Earth and Master It: The Ancient and Medieval Career of a Biblical Text* (Cornell: Cornell University Press, 1989).

Cohen, P. C., *A Calculating People: The Spread of Numeracy in Early America* (Chicago, IL: University of Chicago Press, 1983).

Coughlan, P., 'Natural History and Historical Nature: The Project for a Natural History of Ireland', in Greengrass et al., *Samuel Hartlib*, pp. 298–317.

Cranston, M., *John Locke: A Biography* (New York: Macmillan, 1957).

Crocker, R., *Henry More 1614–1687: A Biography of the Cambridge Platonist* (Dordrecht: Kluwer, 2003).

Cunningham, A., 'How the *Principia* Got Its Name,' *History of Science*, 29 (1991), pp. 377–92.

—, 'The Culture of Gardens', in N. Jardine, J. A. Secord and E. C. Spary (eds), *Cultures of Natural History* (Cambridge: Cambridge University Press, 1996), pp. 38–56.

Daston, L., 'The Moral Economy of Science' *Osiris*, 2nd series, 10 (1995), pp. 2–24.

Davis, J. C., *Utopia and the Ideal Society: A Study of English Utopian Writing 1516–1700* (Cambridge: Cambridge University Press, 1981).

Deacon, M., *Scientists and the Sea 1650–1900: A Study of Marine Science*, 2nd edn (1971; London: Ashgate, 1997).

Dear, P., *Discipline and Experience: The Mathematical Way in the Scientific Revolution* (Chicago, IL: University of Chicago Press, 1995).

Drayton, R., 'Knowledge and Empire', in P. J. Marshall (ed.) *Oxford History of the British Empire, Volume 2: The Eighteenth Century* (Oxford: Oxford University Press, 1998), pp. 231–52.

—, 'Science, Medicine, and the British Empire', in R. W. Winks (ed.) *Oxford History of the British Empire, Volume 5: Historiography* (Oxford: Oxford University Press, 1999), pp. 264–76.

—, *Nature's Government: Science, Imperial Britain and the Improvement of the World* (2000; New Delhi: Orient Longman, 2005).

Drey, T. A. (ed.), *Tacitus* (London: Routledge, 1969).

Dunn, J., *The Political Thought of John Locke* (Cambridge: Cambridge University Press, 1968).

—, 'From Applied Theology to Social Analysis', in *Rethinking Modern Political Theory Essays 1979–83* (Cambridge: Cambridge University Press, 1985).

Eco, U., *The Search for the Perfect Language* (1993; Oxford: Blackwell, 1995).

Elliott, J. H., *The Old World and the New: 1492–1650* (Cambridge: Cambridge University Press, 1970).

—, 'Afterword', in D. Armitage and M. J. Braddick (eds), *The British Atlantic World 1500–1800* (Basingstoke: Palgrave Macmillan, 2002), pp. 233–49.

Feest, C. F., 'The Collecting of American Indian Artifacts in Europe 1493–1750', in K. O. Kupperman (ed.), *America in European Consciousness 1493–1750* (Chapel Hill, NC: University of North Carolina Press, 1995), pp. 324–50.

Ferguson, N., *Empire: How Britain Made the Modern World* (London: Penguin, 2004).

Feuer, L., 'Science and the Ethic of Protestant Asceticism: A Reply to Professor Robert K. Merton', *Research in Sociology of Knowledge, Sciences and Art*, 2 (1979), pp. 1–23.

Findlen, P., 'Courting Nature', in N. Jardine, J. A. Secord and E. Spary (eds), *Cultures of Natural History* (Cambridge: Cambridge University Press, 1996), pp. 57–74.

—, 'Francis Bacon and the Reform of Natural History in the Seventeenth Century', in D. R. Kelley (ed.), *History and the Disciplines: The Reclassification of Knowledge in Early Modern Europe* (Rochester, NY: University of Rochester Press, 1997), pp. 239–60.

—, 'Natural History', in Park and Daston (eds) *The Cambridge History of Science*, pp. 437–71.

Fitzmaurice, A., *Humanism and America: An Intellectual History of English Colonisation 1500–1625* (Cambridge: Cambridge University Press, 2003).

Force, J. E. and R. H. Popkin (eds), *Millenarianism and Messianism in Early Modern European Culture, Volume 3: The Millenarian Turn: The Millenarian Contexts of Science, Politics and Everyday Anglo-American Life in the Seventeenth and Eighteenth Centuries* (Dordrecht: Kluwer Academic Publishers, 2001).

Foster, S., *The Long Argument: English Puritanism and the Shaping of New England Culture, 1570–1700* (Chapel Hill, NC: University of North Carolina Press, 1991).

Foucault, M., *The Order of Things: The Archaeology of the Human Sciences* (New York: Pantheon, 1970).

Gascoigne, J., *Science in the Service of Empire: Joseph Banks, the British State and the Uses of Science in the Age of Revolution* (Cambridge: Cambridge University Press, 1998).

Gaukroger, S., *Francis Bacon and the Transformation of Early Modern Philosophy* (Cambridge: Cambridge University Press, 2001).

Glacken, C. J., *Traces on the Rhodian Shore: Nature and Culture in Western Thought from Ancient Times to the End of the Eighteenth Century* (Berkeley, CA: University of California Press, 1967).

Glausser, W., 'Three Approaches to Locke and the Slave Trade', *Journal of the History of Ideas*, 51 (1990), pp. 199–216.

Golinski, J. V., 'Language, Discourse, and Science', in R. C Olby, G. N. Cantor, J. R. R. Christie and M. J. S. Hodge (eds), *Companion to the History of Modern Science* (London: Routledge, 1990), pp. 110–23.

Greene, R. A., 'Henry More and Robert Boyle on the Spirit of Nature', *Journal of the History of Ideas*, 23 (1962), pp. 451–74.

Greengrass, M., M. Leslie and T. Raylor (eds), *Samuel Hartlib and Universal Reformation: Studies in Intellectual Communication* (Cambridge: Cambridge University Press, 1994).

Grove, R., *Green Imperialism: Colonial Expansion, Tropical Island Edens and the Origins of Environmentalism, 1600–1800* (Cambridge: Cambridge University Press, 1995).

—, *Ecology, Climate and Empire: Colonialism and Global Environmental History: 1400–1940* (Cambridge: Cambridge University Press, 1997).

Hacking, I., *Historical Ontology* (Cambridge, MA: Harvard University Press, 2002).

Hall, M. B., *Henry Oldenburg: Shaping the Royal Society* (Oxford: Oxford University Press, 2002).

Harding, S., *Is Science Multicultural? Postcolonialisms, Feminisms and Epistemologies* (Bloomington, IN: Indiana University Press, 1998).

Harrison, J. and P. Laslett (eds), *The Library of John Locke* (1965; Oxford: Oxford University Press, 1971).

Harrison, P., 'Subduing the Earth: Genesis 1, Early Modern Science, and the Exploration of Nature', *Journal of Religion*, 79 (1999), pp. 86–109.

—, *The Bible, Protestantism and the Rise of Natural Science* (1998; Cambridge: Cambridge University Press, 2001).

—, 'Curiosity, Forbidden Knowledge, and the Reformation of Natural Philosophy in Early Modern England', *Isis*, 92 (2001), pp. 265–90.

—, '"Fill the Earth and Subdue It": Biblical Warrants for Colonization in Seventeenth-Century England', *Journal of Religious History*, 29 (2005), pp. 3–24.

Hening, W., *Statutes at Large, Being a Collection of all the Laws of Virginia, from the First Session of the Legislature in the Year 1619*, 13 vols (1819–23; Charlottesville, VA: University Press of Virginia, 1969).

Henry, J., 'The Scientific Revolution in England', in R. Porter and M. Teich (eds), *The Scientific Revolution in National Context* (Cambridge: Cambridge University Press, 1992), pp. 178–210.

Hill, C., *Intellectual Origins of the English Revolution* (Oxford: Clarendon University Press, 1965).

Hobsbawm, E., *Industry and Empire: From 1750 to the Present Day, Volume 3: The Pelican Economic History of Britain* (1968; Baltimore, MD: Pelican, 1969).

Horn, J., *Adapting to a New World: English Society in the Seventeenth-Century Chesapeake* (Chapel Hill, NC: University of North Carolina Press, 1994).

Hotson, H., 'Philosophical Pedagogy in Reformed Central Europe between Ramus and Comenius: A Survey of the Continental Background of the "Three Foreigners"', in Greengrass et al., *Samuel Hartlib*, pp. 29–50.

Hunter, M., *Science and Society in Restoration England* (Cambridge: Cambridge University Press, 1981).

—, *Establishing the New Science: The Experience of the Early Royal Society* (Woodbridge: Boydell & Brewer, 1989).

—, 'Science and Heterodoxy: An Early Modern Problem Reconsidered', in D. C. Lindberg and R. S. Westman (eds) *Reappraisals of the Scientific Revolution* (Cambridge: Cambridge University Press, 1990), pp. 437–60.

—, *Robert Boyle by Himself and his Friends, with a Fragment of William Wotton's Lost 'Life of Boyle'* (London: Pickering & Chatto, 1994).

— (ed.) *Robert Boyle Reconsidered* (Cambridge: Cambridge University Press, 1994).

—, *Robert Boyle (1627–91), Scrupulosity and Science* (Woodbridge: Boydell & Brewer, 2000).

Hunter, M., and C. Littleton, 'The Work-Diaries of Robert Boyle: A Newly Discovered Source and its Internet Publication', *Notes and Records of the Royal Society*, 55 (2001), pp. 373–90.

Iliffe, R., 'Foreign Bodies: Travel, Empire and the Early Royal Society of London, Part I: Englishmen on Tour', *Canadian Journal of History*, 33 (1998), pp. 357–85.

—, 'Foreign Bodies: Travel, Empire and the Early Royal Society of London, Part II: The Land of Experimental Knowledge', *Canadian Journal of History*, 34 (1999), pp. 23–50.

Ivison, D., 'Locke, Liberalism and Empire', in Anstey (ed.) *The Philosophy of John Locke*, pp. 86–105.

Jacob, J. R., 'Restoration, Reformation and the Origins of the Royal Society', *History of Science*, 13 (1975), pp. 155–76.

—, *Robert Boyle and the English Revolution: A Study in Social and Intellectual Change* (New York: Burt Franklin & Co., 1977).

—, *Henry Stubbe, Radical Protestantism and the Early Enlightenment* (Cambridge: Cambridge University Press, 1983).

—, 'The Political Economy of Science in Seventeenth-Century England', *Social Research*, 59 (1992), pp. 505–32.

Jacob, J. R., and M. Jacob, 'The Anglican Origins of Modern Science: The Metaphysical Foundations of the Whig Constitution', *Isis*, 71 (1980), pp. 251–67.

Josten, C. H., *Elias Ashmole: His Autobiographical and Historical Notes, His Correspondence, and Other contemporary Sources Relating to His Life and Work* (Oxford: Oxford University Press, 1966).

Jowitt, C., 'Colonialism, Jewishness and Politics in Bacon's *New Atlantis*', in Price (ed.) *Francis Bacon's New Atlantis*, pp. 129–55.

Kautsky, K., 'Germany, England and the World Policy', trans. E. Crawford, *The Social Democrat*, 4 (1900), pp. 230–6.

Kellaway, W., *The New England Company, 1649–1776: Missionary Society to the American Indians* (London: Longmans, 1961).

Kelley, D. R. (ed.), *History and the Disciplines: The Reclassification of Knowledge in Early Modern Europe* (Rochester: University of Rochester Press, 1997).

Kingsbury, S. M. (ed.), *The Records of the Virginia Company of London*, 4 vols (Washington, DC: Government Printing Office, 1906–35).

Kneebone, J. T., J. J. Looney, B. Tarter and S. G. Treadway (eds), *Dictionary of Virginia Biography, Volume 2* (Richmond: Library of Virginia, 2001).

Kupperman, K. O., *Roanoke: the Abandoned Colony* (Totowa: Rowman & Allanheld, 1984).

—, *Providence Island: 1630–1641: The Other Puritan Colony* (Cambridge: Cambridge University Press, 1993).

— (ed.), *America in European Consciousness 1493–1750* (Chapel Hill, NC: University of North Carolina Press, 1995).

Lebovic, H., 'The Uses of America in Locke's Second Treatise of Government', *Journal of the History of Ideas*, 47 (1986), pp. 567–81.

Lenin, V., *Imperialism: The Highest Stage of Capitalism* (1917; Moscow: Progress Publishers, 1982).

Leslie, M., and T. Raylor (eds), *Culture and Cultivation in Early Modern England* (Leicester: Leicester University Press, 1992).

Lux, D., and H. J. Cook, 'Closed Circles or Open Networks? Communicating at a Distance during the Scientific Revolution', *History of Science*, 36 (1998), pp. 179–211.

Mabo and Others v. *Queensland* (no. 2) (1992), 175, *Commonwealth Law Reports*, 1.

MacGregor, A., *Tradescant's Rarities: Essays on the Foundation of the Ashmolean Museum, 1683, with a Catalogue of the Surviving Early Collections* (Oxford: Oxford University Press, 1983).

—, *Ark to Ashmolean: The Story of the Tradescants, Ashmole and the Ashmolean Museum* (Oxford: Ashmolean Museum and Tradescant Trust, 1997).

MacGregor, A. G., and A. J Turner, 'The Ashmolean Museum', in L. S. Sutherland and L. G. Mitchell (eds), *The Eighteenth Century, History of the University of Oxford, Volume 5* (Oxford: Clarendon Press, 1986), pp. 639–58.

McGuire, J. E., 'Boyle's Conception of Nature', *Journal of the History of Ideas*, 33 (1972), pp. 523–42.

Mackay, D., *In the Wake of Cook: Exploration, Science and Empire 1780–1801* (New York: St Martin's Press, 1981).

MacKenzie, J. M., *The Empire of Nature: Hunting, Conservation and British Imperialism* (Manchester: Manchester University Press, 1988).

—, *Imperialism and the Natural World* (Manchester: Manchester University Press, 1991).

McClellan, J. E., III, *Science Reorganized: Scientific Societies in the Eighteenth Century* (New York: Columbia University Press, 1985).

MacLeod, R., and M. J. Lewis, *Disease, Medicine and Empire: Perspectives on Western Medicine and the Experience of European Expansion* (London and New York: Routledge, 1988).

Maddison, R. E. W., 'Studies in the Life of Robert Boyle, F.R.S. Part I. Robert Boyle and Some of His Foreign Visitors', *Notes and Records of the Royal Society of London*, 9 (1951), pp. 1–35.

Mandelbrote, S., and J. A. Bennett, 'Biblical Interpretation and the Improvement of Society: Samuel Hartlib and his Circle', *Intellectual News*, 3 (1998), pp. 17–24.

Martin, J., *Francis Bacon, the State and the Reform of Natural Philosophy* (Cambridge: Cambridge University Press, 1992).

Martin, J. F., *Profits in the Wilderness: Entrepreneurship and the Founding of New England Towns in the Seventeenth Century* (Chapel Hill, NC: University of North Carolina Press, 1991).

Marshall, P. J. (ed.), *The Oxford History of the British Empire, Volume 2: The Eighteenth Century* (1998; Oxford: Oxford University Press, 2001).

Maycock, A. L., *Chronicles of Little Gidding* (London: SPCK, 1954).

Merton, R. K., *Science, Technology and Society in Seventeenth-Century England* (1938; New York: Fertig, 1970).

Miller, D. P., and H. Reill (eds), *Visions of Empire: Voyages, Botany and Representations of Nature* (Cambridge: Cambridge University Press, 1996).

Morgan, E. S., *Visible Saints: The History of a Puritan Idea* (Ithaca, NY: Cornell University Press, 1963).

Muir, L. R., and J. A. White (eds), *Materials for the Life of Nicholas Ferrar: A Reconstruction of John Ferrar's Account on his Brother's Life Based on All the Surviving Copies* (Leeds: Leeds Philosophical and Literary Society, 1996).

Mulligan, L., 'Civil War Politics, Religion and the Royal Society', in C. Webster (ed.) *The Intellectual Revolution of the Seventeenth Century* (London and Boston, MA: Routledge & Kegan Paul, 1974), pp. 317–39.

—, 'Puritans and English Science: A Critique of Webster', *Isis*, 71 (1980), pp. 457–69.

Muthu, S., *Enlightenment Against Empire* (Princeton, NJ: Princeton University Press, 2003).

Newman, W. R., *Gehennical Fire; the Lives of George Starkey, an American Alchemist in the Scientific Revolution* (1994; Chicago, IL: University of Chicago Press, 2002).

O'Brien, J. J., 'Samuel Hartlib's Influence on Robert Boyle's Scientific Development Part II: Boyle in Oxford', *Annals of Science*, 21 (1965), pp. 257–76.

Osler, M. J., 'From Immanent Natures to Nature as Artifice: The Reinterpretation of Final Causes in Seventeenth-Century Natural Philosophy', *The Monist*, 79 (1996), pp. 388–408.

Palladino, P., and M. Worboys, 'Science and Imperialism', *Isis*, 84 (1993), pp. 91–102.

Parekh, B., 'Liberalism and Colonialism: A Critique of Locke and Mill', in J. Pieterse and B. Parekh (eds), *The Decolonization of the Imagination: Culture, Knowledge, and Power* (London: Zed Books, 1995), pp. 81–98.

Park, K., and L. Daston (eds), *The Cambridge History of Science, Volume 3: Early Modern Science* (Cambridge: Cambridge University Press, 2006).

Peacey, J., 'Seasonable Treatises: A Godly Project of the 1630s', *English Historical Review*, 113:452 (June 1998), pp. 667–9.

Peck, L. L. (ed.), *The Mental World of the Jacobean Court* (Cambridge: Cambridge University Press, 1991).

Peltonen, M., 'Politics and Science: Francis Bacon and the True Greatness of States', *Historical Journal*, 35 (1992), pp. 279–305.

—, *Classical Humanism and Republicanism in English Political Thought 1570–1640* (Cambridge: Cambridge University Press, 1995).

— (ed.) *The Cambridge Companion to Bacon* (Cambridge: Cambridge University Press, 1996).

Perez-Ramos, A., *Francis Bacon's Idea of Science and the Maker's Knowledge Tradition* (Oxford: Clarendon Press, 1988).

—, 'Bacon's Forms and the Maker's Knowledge Tradition' in Peltonen (ed.) *Cambridge Companion to Bacon*, pp. 99–120.

Petitjean, P., C. Jami and A. M. Moulin (eds), *Science and Empires: Historical Studies about Scientific Development and European Expansion* (Dordrecht: Kluwer Academic Publishers, 1991).

Poovey, M., *A History of the Modern Fact: Problems of Knowledge in the Sciences of Wealth and Society* (Chicago, IL: University of Chicago Press, 1998).

Prest, J., *The Garden of Eden: The Botanic Garden and the Re-creation of Paradise* (New Haven, CT: Yale University Press, 1981).

Price, B. (ed.), *Francis Bacon's New Atlantis: New Interdisciplinary Essays* (Manchester: Manchester University Press, 2002).

Principe, L. M., 'Boyle's Alchemical Pursuits', in Hunter (ed.) *Robert Boyle Reconsidered*, pp. 91–105.

Privratska, J., and V. Privratsky, 'Language as the Product and Mediator of Knowledge: The Concept of J. A. Comenius', in Greengrass et al., *Samuel Hartlib*, pp. 162–73.

Quinn, D. B., 'Ireland and Sixteenth-Century European Expansion', in T. D. Williams (ed.) *Historical Studies: Papers Read before the Irish Conference of Historians*, 1 (London: Bowes & Bowes, 1958), pp. 20–32.

Reingold, N., and M. Rothenberg (eds), *Scientific Colonialism: A Cross-Cultural Comparison* (Washington, DC: Smithsonian Institution Press, 1987).

Reynolds, H., *The Law of the Land*, 2nd edn (Melbourne: Penguin, 1992).

Rich, E. E. (ed.), *Minutes of the Hudson's Bay Company 1679–1684, First Part 1679–1682* (Toronto: Champlain Society, 1945).

Rose, J. H., A. P. Newton and E. A. Benians (eds), *The Cambridge History of the British Empire, Volume 1: The Old Empire from the Beginnings to 1783* (Cambridge: Cambridge University Press, 1929).

Rossi, P., *Francis Bacon: From Magic to Science*, trans. S. Rabinovitch (London: Routledge, 1968).

—, 'Bacon's Idea of Science' in Peltonen (ed.) Cambridge *Companion to Bacon*, pp. 25–46.

Rusnock, A., 'Biopolitics: Political Arithmetic in the Enlightenment', in W. Clark, J. Golinski and S. Schaffer (eds), *The Sciences in Enlightened Europe* (Chicago, IL: University of Chicago Press, 1999), pp. 49–68.

Sacks, D. H., 'Rebuilding Solomon's Temple: Richard Hakluyt and Empire in the "Age of Discovery"', unpublished conference paper.

—, *The Widening Gate: Bristol and the Atlantic Economy, 1450–1700* (Berkeley, CA: University of California Press, 1991).

Said, E., *Culture and Imperialism* (New York: Alfred A. Knopf, 1993).

Sainty, J. C. (ed.), *Officials of the Board of Trade, 1660–1870* (London: Athlone Press, 1974).

Salisbury, N., *Manitou and Providence: Indians, Europeans, and the Making of New England, 1500–1643* (New York: Oxford University Press, 1982).

Schaffer, S., 'Golden Means: Assay Instruments and the Geography of Precision in the Guinea Trade', in M.-N. Bourguet, C. Licoppe and H. O. Sibum (eds), *Instruments, Travel and Science: Itineraries of Precision from the Seventeenth to the Twentieth Century* (London and New York: Routledge, 2002), pp. 20–50.

Seed, P., *Ceremonies of Possession: Europe's Conquest of the New World 1492–1640* (Cambridge: Cambridge University Press, 1995).

Seeley, J. R., *The Expansion of England* (London: Macmillan, 1883).

Serjeantson, R., 'Natural Knowledge in the *New Atlantis*', in Price (ed.) *Francis Bacon's New Atlantis*, pp. 82–105.

Shanahan, T., 'God and Nature in the Thought of Robert Boyle', *Journal of the History of Philosophy*, 26 (1988), pp. 547–69.

Shapin, S., 'Pump and Circumstance: Robert Boyle's Literary Technology', *Social Studies of Science*, 14 (1984), pp. 481–520.

—, 'The House of Experiment in Seventeenth-Century England', *Isis*, 79 (1988), pp. 373–404.

—, *A Social History of Truth: Civility and Science in Seventeenth-Century England* (Chicago, IL: University of Chicago Press, 1994).

Shapin, S., and S. Schaffer, *Leviathan and the Air Pump: Hobbes, Boyle and the Experimental Life* (Princeton, NJ: Princeton University Press, 1985).

Shapiro, B., 'Latitudinarianism and Science in Seventeenth-Century England', *Past and Present*, 40 (1968), pp. 16–41.

—, *A Culture of Fact: England 1550–1720* (Ithaca, NY: Cornell University Press, 2000).

Stimson, D., 'Hartlib, Haak and Oldenburg: Intelligencers', *Isis*, 31 (1940), pp. 309–26.

—, *Scientists and Amateurs: A History of the Royal Society* (New York: H. Schuman, 1948).

Swann, M., *Curiosities and Texts: The Culture of Collecting in Early Modern England* (Philadelphia, PA: University of Pennsylvania Press, 2001).

Thackray, A. (ed.), *Constructing Knowledge in the History of Science* (Chicago, IL: University of Chicago Press, 1995).

Thompson, P., 'William Bullock's Strange Adventure: A Plan to Transform Seventeenth-Century Virginia', *William and Mary Quarterly*, 61 (2004), pp. 107–28.

Trevor-Roper, H., *Religion, the Reformation and Social Change, and Other Essays*, 2nd edn (London: Macmillan, 1967).

Tully, J., *A Discourse on Property: John Locke and His Adversaries* (Cambridge: Cambridge University Press, 1980).

—, *An Approach to Political Philosophy: Locke in Contexts* (Cambridge: Cambridge University Press, 1993).

Turnbull, G. H., *Samuel Hartlib: A Sketch of His Life and His Relations to J. A. Comenius* (Oxford: Oxford University Press, 1920).

—, *Hartlib, Dury and Comenius: Gleanings from Hartlib's Papers* (Liverpool: University of Liverpool Press and Hodder & Stoughton, 1947).

Ullmann, W., 'This Realm of England is an Empire', *Journal of Ecclesiastical History*, 30 (1979), pp. 175–203.

Vickers, B., 'Francis Bacon and the Progress of Knowledge', *Journal of the History of Ideas*, 53 (1992), pp. 495–518.

Waldron, J., *God, Locke and Equality: Christian Foundations of Locke's Political Thought* (Cambridge: Cambridge University Press, 2002).

Webster, C., *Samuel Hartlib and the Advancement of Learning* (Cambridge: Cambridge University Press, 1970).

—, 'The Authorship and Significance of *Macaria*', *Past and Present*, 56 (1972), pp. 34–49.

— (ed.), *The Intellectual Revolution of the Seventeenth Century* (London and Boston, MA: Routledge & Kegan Paul, 1974).

—, 'New Light on the Invisible College: The Social Relations of English Science in the mid Seventeenth Century', *Transactions of the Royal Historical Society*, 24 (1974), pp. 19–42.

— (ed.), *Utopian Planning and the Puritan Revolution: Gabriel Plattes, Samuel Hartlib and Macaria* (Oxford: Wellcome Unit for the History of Medicine, 1979).

—, 'Benjamin Worsley: Engineering for Universal Reform from the Invisible College to the Navigation Act', in Greengrass et al., *Samuel Hartlib*, pp. 213–35.

—, *The Great Instauration: Science, Medicine and Reform, 1626–1660*, 2nd edn (1975; Bern: Peter Lang, 2002).

Weinberger, J., *Science, Faith and Politics: Francis Bacon and the Utopian Roots of Modern Age, a Commentary on Bacon's Advancement of Learning* (Ithaca, NY: Cornell University Press, 1985).

White, H. B., 'Bacon's Imperialism', *American Political Science Review*, 102 (1958), pp. 470–89.

—, *Peace Among the Willows: The Political Philosophy of Francis Bacon* (The Hague: Martinus Nijhoff, 1968).

White, L., 'The Historical Roots of Our Environmental Crisis', *Science*, 155 (1967), pp. 1202–7.

Whitney, C., *Francis Bacon and Modernity* (New Haven, CT: Yale University Press, 1986).

—, 'Francis Bacon's *Instauratio:* Dominion of and over Humanity', *Journal of the History of Ideas*, 50 (1989), pp. 371–90.

—, 'Merchants of Light: Science as Colonization in the *New Atlantis*', reprinted in W. Sessions (ed.) *Francis Bacon's Legacy of Texts* (New York: AMS Press, 1990), pp. 255–68.

Williams, A., *The Common Expositor: An Account of the Commentaries on Genesis 1527–1633* (Chapel Hill, NC: University of North Carolina Press, 1948).

Williams, R., *Keywords: A Vocabulary of Culture and Society* (1976; Oxford: Oxford University Press, 1992).

Wilson, B., *The Hive: The Story of the Honeybee and Us* (London: John Murray, 2004).

Winslow, O. E., *John Eliot, 'Apostle to the Indians'* (Boston, MA: Houghton Mifflin, 1968).

Worboys, M., 'Science and British Colonial Imperialism, 1895–1940' (unpublished D.Phil. thesis, University of Sussex, 1980).

Yates, F. A., *Astraea: The Imperial Theme in the Sixteenth Century* (London: Routledge, 1975).

—, *Giordano Bruno and the Hermetic Tradition* (1964; London: Routledge 2002).

—, *The Rosicrucian Enlightenment* (1972; London: Routledge, 2002).

Yeo, R., *Encyclopaedic Visions: Scientific Dictionaries and Enlightenment Culture* (Cambridge: Cambridge University Press, 2001).

Young, J. T., *Faith, Medical Alchemy and Natural Philosophy: Johann Moriaen, Reformed Intelligencer, and the Hartlib Circle* (Aldershot: Ashgate, 1998).

Young, R. F., *Comenius and the Indians in New England* (London: King's College, 1929).

—, *Comenius in England* (London: Oxford University Press, 1934).

INDEX